DATE DUE

TOXIN
The Cunning of Bacterial Poisons

TOXIN

The Cunning of Bacterial Poisons

ALISTAIR J. LAX

OXFORD
UNIVERSITY PRESS

OXFORD

UNIVERSITY PRESS

Great Clarendon Street, Oxford OX2 6DP

Oxford University Press is a department of the University of Oxford.
It furthers the University's objective of excellence in research, scholarship,
and education by publishing worldwide in

Oxford New York

Auckland Cape Town Dar es Salaam Hong Kong Karachi
Kuala Lumpur Madrid Melbourne Mexico City Nairobi
New Delhi Shanghai Taipei Toronto

With offices in

Argentina Austria Brazil Chile Czech Republic France Greece
Guatemala Hungary Italy Japan Poland Portugal Singapore
South Korea Switzerland Thailand Turkey Ukraine Vietnam

Oxford is a registered trade mark of Oxford University Press
in the UK and in certain other countries

Published in the United States
by Oxford University Press Inc., New York

British Library Cataloguing in Publication Data

Data available

Library of Congress Cataloging in Publication Data

Lax, Alistair J., 1953–
Toxin : the story of bacterial poisons / Alistair J. Lax
p. ; cm.
Includes bibliographical references and index
ISBN–13: 978–0–19–860558–4 (alk. paper)
ISBN–10: 0–19–860558–7 (alk. paper)
1. Bacterial toxins—Popular works. I. Title
[DNLM: 1. Bacterial toxins—history. 2. Bacterial toxins
—poisoning. QW 11.1 L425t 2005]
QP632.B3L39 2005
615.9′5293—dc22
2005019702

Typeset by Footnote Graphics Limited
Printed in Great Britain
on acid-free paper by
Clays Ltd., St. Ives plc

ISBN 0–19–860558–7 978–0–19–860558–4

1

Preface

I became interested in bacterial toxins around 15 years ago, and was immediately and increasingly fascinated as I learnt more about how these ingenious bacterial poisons take over control of our cells. Although toxins are made in bacteria, they are designed specifically to work inside our cells. There each toxin can identify with precision a key function that makes the cell work, and moreover interfere with its normal function. Toxins therefore possess an intimate knowledge of our cells, and this is the reason why bacteria are such lethal molecules and cause disease. Toxins remain a significant cause of disease today, not just in the developing world, where they are a major cause of death, but also in the west. New toxin-linked diseases are still appearing and we are still finding new ways that these poisons interact with us.

The story of toxins and of the scientists who worked on them is one of genius, endeavour, and personal squabbles. It is also entwined with the birth of modern biology in the last two centuries. Inspirational and eccentric characters fashioned the new science of microbiology by the late nineteenth century and this in turn led to other new sciences, such as immunology and virology.

The potent action of bacterial toxins can be harnessed for good or evil. Not surprisingly they have been used as weapons by unscrupulous regimes and terrorist groups. We can also use information about their poisonous nature to fight bacterial disease by making vaccines, and we can even turn the toxic activity to beneficial use in novel ways to try to fight cancer. In addition, as toxins interact so precisely with our cells, they can be used to help us to understand cell function in health and disease. What wonderfully interesting molecules they are!

Alistair Lax
April 2005

Acknowledgements

It has been said that copying from one source is plagiarism, whereas copying from more than one is research. On that basis I hope that this book will be regarded as a work of research. In some sections of this book I have had the opportunity to look up several sources including original published papers. In other places I have relied heavily on solitary or limited resources, whereas the source of information on other aspects of toxins comes from attendance at conferences and individual contacts. I have tried to acknowledge the written sources in a section at the end of this book, and mention various direct contacts here. I am aware that neither list is entirely comprehensive and I apologise in advance for any especially glaring omissions. Of course any errors of fact in the book are entirely my responsibility.

Many people provided information, or suggested where I could find it. I would particularly like to thank Bob Arnott, John Blair, Tim Carter, Ralph Ferichs, John Forrester, Ian Reader, Katie Sambrook, and Rick Titball.

I would like to thank the following people for generously providing images that I could use: John Collier of Harvard University, Ignatius Ding of the Alliance for Preserving the Truth of Sino-Japanese War, Ute Hornbogen at the Robert Koch Museum, Robin Keeley of the Forensic Science Service, and Tim Staten of Wolverhampton Archives and Local Studies.

Several members of my family and several friends were kind enough to read early drafts and give me their helpful and candid comments. I would like to thank Alan Hester, Pauline Lax, Nick Lax, Isobel Lax, and Hugh Smith.

Finally, I thank my wife Pauline, and Nick, Tom and Ollie for their forbearance, support and love over the many years I took to write this book.

Contents

Figures and Tables

1

TOXINS ARE EVERYWHERE
How toxins have affected history

The Bulgarian playwright and novelist Georgi Markov was a marked man. He had defected to Britain in 1971 and was hated by the uncompromising communist junta that ruled his native country for his outspoken broadcasts against them. It was Thursday 7 September 1978, the birthday of Todor Zhirkov, the Bulgarian leader, when the fatal blow was struck. It was at least the third attempt to kill him, and after the previous failed attack he had heard that the next attempt would be 'special'. Markov was standing at his usual bus stop on Waterloo Bridge on his way to the BBC for his afternoon broadcast. He felt a sharp pain at the back of his right thigh. When he turned round, he saw a man picking up an umbrella. The man muttered an apology then hailed a taxi. He did this with some difficulty, because Markov noted that the taxi driver did not appear to understand him, suggesting that he was not local. At work, Markov's leg was now very painful, and he showed a colleague what looked like a bee sting with a puncture mark. The next day he became very ill, with fever and vomiting. That night he went to St James's Hospital in Balham, in south London. Within three days he was dead.

The doctors had been puzzled by his symptoms when he had arrived at the hospital. Attempts to find a bacterial infection failed, and X-rays of his affected thigh did not reveal anything abnormal. Over the next few days his white blood cell count rose dramatically, suggesting that his body was trying to fight some type of infection or poison. On Saturday, just over two days after the attack, his blood pressure collapsed and he began passing blood in his urine. He was also vomiting blood and by this time was in intensive care. On Sunday morning his heart stopped. It was clear that something very unusual had killed him. A postmortem examination was carried out the following day by Dr Rufus Crompton, the Home Office Pathologist, and sections from each thigh were taken for more detailed examination. Crompton said that many had initially thought that Markov's story of the attack on him was unlikely because of his known paranoia about the Bulgarians.

The postmortem examination showed that Markov's body had suffered a devastating attack of some kind. The doctors immediately thought of poisons, but were unsure what type could have been used. The tissue removed from his leg was sent to the British scientific defence laboratories at Porton Down and was examined by Dr David Gall. There a small pellet, 1.53 mm in diameter (the precision of the measurement turned out to be important), was found on the tissue. The discovery of the pellet was serendipitous, as Gall initially thought that a tiny metallic speck on the surface of the tissue was a pin that had been pushed in during the postmortem examination. When he idly touched it with his finger it moved across the tissue and he realised that it might have some significance. The pellet could be seen on subsequent more careful analysis of the X-rays that had been taken at Markov's admission to hospital.

Two holes had been drilled at right angles in the pellet and each was 0.34 mm in diameter (Figure 1.1). The pellet was sent to the Metropolitan Police Forensic Science Laboratory where Dr Robin Keeley looked at it. He found out that it was made of platinum and iridium, an extremely hard alloy that would not have deformed on passing through Markov's clothing. It was so hard that special equipment must have been used to drill the holes. In addition, this alloy is biologically inert, so it would not have announced its presence by irritating the body where it had lodged. Keeley also stated that there was no evidence of the pellet being delivered by an explosive device, so it is thought likely that it had been delivered by a gas gun in the adapted umbrella.

Porton Down could not find any poison in Markov's body, so the scientists had to work on the symptoms of his illness, and the miniscule pellet that had delivered its deadly cargo. The volume of the two holes drilled in the pellet was in the order of 0.3 mm^3, or 3 ten-millionths of a litre. This tiny volume could contain about 0.5 mg of material, less than a thousandth of a gram. It appeared unlikely

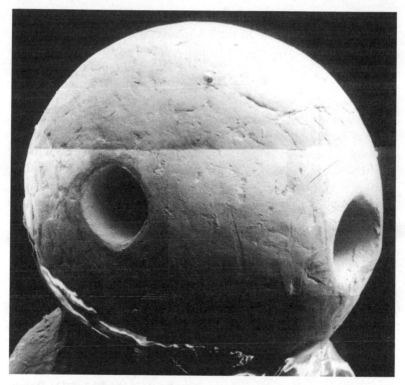

1.1 The pellet that killed Georgi Markov—composite of photographs taken with a scanning electron microscope.
(Provided by Robin Keeley of the Forensic Science Service, UK.)

that his death had been caused by an infectious agent—death was too rapid and there was no evidence of a local infection. Few chemical poisons were toxic enough to have killed with the tiny quantity that the pellet could hold, and those that could did not fit with the symptoms. It appeared likely that a much more deadly type of poison had been used—a biological poison or toxin. Indeed few biological toxins were toxic enough, so they considered whether any of the most dangerous ones had been responsible. The symptoms did not suggest the bacterial toxins diphtheria, tetanus, or botulinum, but they did fit with the plant toxin ricin. Porton Down had experience with ricin for use as a biological weapon, but they had looked at its effects only on small animals. They decided to inject a pig with ricin and found that the affected animal showed similar symptoms to Markov's. Markov had been attacked with one of Nature's most deadly weapons, a biological toxin.

Toxins are powerful biological poisons that are released by bacteria and some plants, as well as snakes and some marine life. This book is the story of bacterial toxins, although I have included ricin because it behaves exactly like a bacterial toxin. The concept that bacteria can produce poisons is easy to understand. However, many of these toxins display a surprising subtlety in their method of attack. They are not crude weapons that mount an indiscriminate onslaught on our bodies, but display an extraordinary intricacy. Toxins are cleverly designed molecules that are manufactured in one type of life form for use in another, and they act as if they have a detailed knowledge of how our bodies work. In fact the smartest of them do not immediately kill the cell that they are attacking, but instead instruct it how to behave for the benefit of the bacterium that sent the toxin there in the first place.

Toxins cause the main symptoms of disease for many dangerous bacteria. Numerous diseases caused by toxins are well known and still feared. These include tetanus, diphtheria, cholera, typhoid, and the plague. Children are immunised against the first two, whereas cholera and typhoid make their unwelcome appearance every year during flooding and famine in the developing world. The plague is part of our history—'Ring-a-ring o' roses, a pocket full of posies, Atishoo atishoo, we all fall down' is not a sweet little nursery rhyme, but refers to the symptoms of plague, a supposed remedy (carrying flowers) and the sneezing that was often the first sign of deadly plague.

In the west, these diseases may seem historical, like a morbid roll call of the appalling battles of the First World War. However, we do not need to go back too far in our own history to find these diseases as major killers. Although we now have means to combat many toxin diseases, it is a disgrace that many of them remain killers in the developing world. Yet no one in the relatively cosy position of the western world should be complacent about the continuing ability of these bacteria to cause harm or even kill us.

The widely hailed demise of bacterial infection in the 1960s, mainly fuelled by the success of antibiotics, was a woefully extravagant and premature claim. It is widely alleged[1] that the then US Surgeon General, William H Stewart stated in 1969:

We can close the book on infectious diseases.

Certainly hindsight has shown this to be a remarkable viewpoint. Even leaving aside the diseases caused by viruses that were not curable then or now, there were many bacterial diseases still causing havoc then, and the simple and obvious fact is that bacterial diseases never did disappear. They remain highly significant, particularly in the developing world. A few statistics from the World Health Organization (WHO) make this very clear. Diarrhoea caused by *Shigella* species is estimated to kill 600,000 people each year, and the same number die from typhoid. Whooping cough kills several hundred thousand people, mainly young children, each year. Tetanus in newborn babies kills about a quarter of a million worldwide each year. This is a truly shocking situation because many of these deaths could be prevented by vaccine programmes and/or improved hygiene. There have been major cholera epidemics in Asia within the last 50 years. Indeed, cholera continues to cause alarm when famine, war, or natural disasters lead to contaminated water supplies. Thousands died of cholera in the civil war in Bangladesh in 1971 and thousands more in the refugee camps of Rwanda in 1994.

Immunisation and antibiotics may have slowed the infectious carnage in the west, but bacterial diseases caused by toxin-producing bacteria are still a justifiable cause for public concern. Many infectious illnesses have only been recently recognised, such as Legionnaire's disease and *Escherichia coli* O157. A surprise for many was the infectious nature of stomach ulcers and stomach cancers—linked to infection with the bacterium *Helicobacter pylori*. Moreover there has been a rapid and worrying rise in antibiotic resistance, most prominently underlined by the concern about MRSA (methicillin-resistant *Staphylococcus aureus*). Gastroenteritis induced by *Salmonella* species remains a perennial problem. That deadly disease, diphtheria, caused alarm in the 1990s when immunisation programmes began to fall apart as the newly independent states formed from the former Soviet Union.

This on-going slaughter by these clever poisons is reason enough to promote research that can lead to effective therapy against their action. This is particularly so in the developing world where there are real constraints on the types of vaccine that can be delivered: for example, cost, the need for special storage in fridges or deep freezers, and an obvious preference for vaccines that can be taken orally, thus avoiding the use of needles. The case for work on bacterial infections is as strong as it ever was.

It is not just through the shadowy world of espionage and assassinations that toxins have affected history. That terrible trinity of famine, warfare, and disease has been linked since antiquity. Plagues of one sort or another contributed to both the fall of Athens and the death throes of the Roman Empire, although it is not clear what the diseases were. Many infectious diseases come from animals, initially linked to early man's decision to associate with animals, exposing him to a new range of bacteria and viruses. Such diseases include anthrax from animal skins, plague from rats, *Salmonella* food poisoning from cattle and poultry, and more recently *E. coli* O157 from cattle. Although these microbes often coexisted harmlessly in their animal host, they were able to take advantage of a new human host that was not equipped to deal with them, and thus offered new opportunities. This feature, coupled with the close and unhygienic living that is a result of urbanisation and compounded by increased travel, has been responsible for most plagues during the centuries. Moreover, the succession of pestilences shortly after the life of Christ is thought to have led to the rise in the power of the Christian Church. The Church promised salvation from disease and death, at the price of linking disease as a punishment for sin. This also gave the Church a stranglehold on Science and Medicine, which it was to wield with varying degrees of malevolent domination for about 1000 years.

In recorded history one disease stands out that wrought havoc and terror on an unprecedented scale and drastically affected history more than any other—the plague. It is not of course the only infectious disease to have killed millions. The pandemic* caused by Spanish flu (a viral disease) in 1918, which resulted in a death toll estimated to be at least 20 million, did not greatly affect history. The bacterial disease typhus, not a toxin-linked disease, was linked to warfare for

---

*A pandemic is a disease that spreads to affect the world. An epidemic is more localised.

centuries and helped to defeat Napoleon in 1812. It also killed millions in Russia after the First World War. The long-term effects of AIDS (acquired immune deficiency syndrome), which is caused by the human immunodeficiency virus (HIV), are currently impossible to predict, although of course the devastating effect on several African societies is already becoming obvious. However, none of these dreadful diseases can yet be put in the same category as the plague. It was known as the 'Black Death' in the fourteenth century, and in its earlier appearance in the sixth century as the 'plague of Justinian'. Much has been written about the Black Death. The plague of Justinian is so far less widely discussed, although it appears to have been as dreadful as the Black Death. After the plague of Justinian the known world moved into the Dark Ages, while the Black Death led on to the Renaissance, and it is thought by many that these terrible pandemics played a significant part in the direction of world history at these times.

The plague, true plague, as opposed to the use of the word as a general term for a widespread dreadful disease, has swept round the world in three terrifying pandemics. There have been only three different types of the plague bacillus,* one linked to each of the three vast world pandemics that began in the 540s, 1340s, and 1890s. Each pandemic was followed by several smaller epidemics. The loss of life exacted by each pandemic is impossible to calculate with any accuracy, but it was huge. One estimate puts the death toll caused by plague at about 200 million over the centuries. Even the death toll in the last pandemic at the end of the nineteenth century is not known precisely, but is estimated at about 10 million. As a percentage of the population this is obviously less than the 40–50 per cent annihilation caused by the earlier pandemics.

The plague of Justinian appeared in AD 541 and reappeared over the next 50–100 years and perhaps longer. This plague took its name from the Byzantine ruler of the Eastern Empire. Its origin may have been African, but it clearly affected Egypt and spread through Palestine and into Constantinople, where it killed about 40 per cent of the population. So many died that there was nowhere to bury them. Roofs were taken off the city's towers to pile the dead in and ships

*A bacterium (plural bacteria) that is rod shaped. Round bacteria are called cocci.

filled with bodies were sent drifting out to sea. From Constantinople the plague spread across Europe throughout the former Roman Empire, possibly killing half the European population. It spread through other parts of the civilised world, through North Africa and India, and into China.

This catastrophe led to a move away from crowded city life and temporarily halted trade. Social structure broke down and people were terrified and desperate. People abandoned husbands or wives who showed the first symptoms of plague in an attempt to escape its clutches. It has been argued that this plague led, at least in part, to the downfall of the whole structure of the Roman Empire and helped precipitate its eventual end. It was viewed as further evidence of sin, and the Church was still seen to offer the only hope of salvation, if not in this life then in the next. In Britain the weakened social structure allowed the Saxon invasion. In the Middle East and northern Africa, it is possible that the destruction caused by the plague eased the way for the Muslim advances there.

Almost exactly 800 years later the next wave of the plague, referred to as the Black Death, came from China in the 1340s, first to the Crimean peninsula by 1346 and then it swept across Europe. The Crimean city of Kaffa, now the city of Feodossia in the Ukraine, was under siege from the Tartars. It is believed that the people of Kaffa caught the plague from the infected corpses that were catapulted into the city by the retreating Tartar troops, whose own army was being ravaged by the plague. The Tartars carried the disease to Russia and India. Traders from Genoa, who had been trapped in the city, then carried the disease on to Italy. It first reached England in June 1348. It arrived in London in November, died down over the winter and then began to increase again in the warmer spring, spreading to the rest of England. The Scots might have escaped but for their ill-founded decision to attack a weakened England in 1350. The Scottish army became infected and spread the disease throughout Scotland on their return home. The Black Death is estimated to have killed about a third of the European population. Some calculations go as high as 45 per cent, although there are not any reliable figures. Certainly, if the tolls from the successive waves of disease are added together, a death rate of greater than 50 per cent is probably not an over-estimate. Elsewhere death and destruction were as terrible: North

Africa, the Middle East, India, and China all lost an unknown but vast proportion of their population.

The onset of the Black Death ushered in a period of 300 years of plague in Europe. Plague during these years showed varying degrees of severity. Several successive waves hit Europe before the end of the fourteenth century, of which the next most deadly was 13 years later in 1361. This outbreak appeared particularly to affect children, perhaps not surprisingly because many of the adult population would either be survivors from 1348 with immunity or people who were naturally immune.

The Black Death affected the social and economic structure of the remaining society. The portrayal of the plague by the Pope's doctor, Guy de Chauliac, again highlights the terrifying social implication of the terror, where:

The father did not visit his son, nor the son his father.
Charity was dead and hope destroyed.[2]

Sienna in Italy was an important trading city at that time, with a population around 40,000—comparable to the then population of London. The City Fathers had begun to build a vast cathedral when the plague struck. One great wall had been completed, but work stopped. By the time the epidemic had ravaged the city, the population had dwindled to 15,000 and it has not returned to the pre-plague level in the next 650 years. Not only was there no longer much enthusiasm for cathedral building, but there was no need for such a large one. When the city eventually got round to building a cathedral a much smaller building was constructed. The original wall still stands alone, perhaps, as Rino Rappuoli* has suggested, as a monument to infectious diseases.

In England the plague had a big impact on society. Plague attacked sporadically. It could wipe out all the labourers on one farm and not kill any on the neighbouring estate. This led to higher wages to tempt workers to move. Such mobility had been unheard of, because peasants were tied by the feudal system to an individual landowner. The English Government had to react to try to restore stability and prevent economic recession. However, despite the failure of the Peasants' Revolt in 1381, the feudal system would never be the same

---

*Rino Rappuoli works in Sienna on vaccine development, particularly on typhoid, and has championed the development of vaccines for developing countries.

and it failed within about another 100 years, allowing England to develop and expand its influence. In addition, the emphasis of farming changed to take account of the reduced labour force. There was a move to tenant farms. There was also less labour-intensive arable farming, and more animals were kept. The plague also had a brief effect on women's emancipation. The shortage of workers led to women being brought into the workforce, and their capacity to earn their own wages led naturally to greater independence. Although the Black Death wrought havoc, its long-term effect on European society was not as negative as the plague of Justinian, and in the aftermath came a period of artistic and scientific rebirth—the Renaissance.

One noticeable effect of the Black Death was on the Christian Church in Europe. Of course the Church still promoted the view, still espoused by some religious people today, that illness was directly linked to sin. An extreme expression of this stance was the growth in the flagellant movement who conducted both public and private ritualistic floggings as a means of purging sin. The flagellants had been around before the Black Death, but grew both in number and in the extent of their organisation. Although they were initially welcomed by the Church, the Pope later turned against the movement when it threatened to challenge his established church. In addition, the need to apportion blame led to the persecution of several groups, in particular the Jews. For some it appeared that the Jewish people suffered less from the pestilence, although it is clear that their communities suffered too. Despite edicts from the Pope, thousands of Jews were killed in Switzerland and in many towns in central Europe. The Strasbourg massacre has been particularly well documented. Following this pogrom, many Jewish people migrated to settle in high numbers in Poland, where there was less persecution and less plague.

On the other hand, the Black Death also led to the diminution of the power and influence of the established Christian Church. In part this was because of the replacement of established priests killed by the plague with untrained men who were not interested in their parishioners.* In part it reflected a growing lack of respect for the mighty organisation that had proved impotent in dealing with this terrible

*The great shortage of priests also led to the founding of university colleges to re-populate the clergy. Corpus Christi, Cambridge and New College, Oxford were among those founded directly as a consequence of the Black Death.

disease. The Church's enthusiasm for connecting disease with moral transgressions was becoming less acceptable. Furthermore, the great likelihood of imminent death during the plague led to a general hopelessness, and thus to a more hedonistic live-for-the-day lifestyle. Boccaccio's great work, the *Decameron*, is a series of stories told by different members of an imaginary cast. In the preface, Boccaccio describes in some detail the effects of the plague on the breakdown of society. Although some people avoided all excesses, others:

Maintained, that to drink freely, frequent places of public resort, and to take pleasure with song and revel, sparing to satisfy no appetite, and to laugh and mock at no event was the sovereign remedy for so great an evil ...[3]

During the Black Death many thought that human civilisation might end. It is hard for us, even as we hear of the ravages brought by AIDS to African states, to imagine the fourteenth-century horror of dead towns, rotting corpses, the breakdown of law and order, the ending of civilisations, and the bleak despair of people who saw no future for themselves, their children, their own people, or even the human race. As the Italian poet Petrarch said at the time[4]:

Is it possible that posterity can believe these things? For we, who have seen them, can hardly believe them.

The recurrence of the plague throughout the next three centuries continued to terrify and shape society. The plague that hit London in 1592–1593 led to many leaving the city. Although the theatres closed, Shakespeare remained in London and turned his great talent from writing plays to writing sonnets. In Spain, waves of plague in the sixteenth and seventeenth centuries probably led to its reduction in world status. The Great Plague in the 1660s was chronicled by well-known writers such as Samuel Pepys and Daniel Defoe. Defoe's book, *A Journal of the Plague Year*, details his personal memoirs of this time. Defoe recounts the horror and agony of plague victims:

The swellings which were generally in the Neck, or Groin, when they grew hard, and would not break, grew so painful, that it was equal to the most exquisite Torture; and some not able to bear the Torture threw themselves out of Windows, or shot themselves, ... [some] vented their Pain by incessant Roarings, and such loud and lamentable Cries were to be heard as we walk'd along the Streets, that would Pierce the very Heart to think of ...

The Great Plague was the last substantial European outbreak of the plague. The reasons for its European disappearance are not clear. Certainly no intentional human intervention can be claimed. One theory is that the black rat was specifically responsible for spreading plague and that it was killed off by the fiercer brown rat, which appeared in Europe in the 1700s and was less likely to act as a carrier. This has not been proved.

The last pandemic started in China in the 1890s, where reports spoke of over 100,000 dead. It spread to the Chinese district of Hong Kong in 1894, later spreading worldwide, including Europe, with a death toll in the millions. It was particularly bad in India and China. It is believed that the Americas were free of plague until the disease struck San Francisco in 1900. To this day plague still exists in pockets around the world, causing around 2,000 deaths each year, and remains a disease of concern in developing countries. The plague can now be treated with antibiotics, and the rats and fleas that are part of the plague cycle can also be targeted. It is nevertheless still a very dangerous disease and deaths occur. Epidemics remain a possibility, as in India in 1994.

As well as its effect on the development of human history, plague also selected people who were more resistant to it. The plague pandemics may have affected two other diseases—leprosy and AIDS. Leprosy patients were more or less eliminated by the Black Death and it is certainly true that leprosy does not exist in parts of the world that have been previously devastated by the plague. More controversially, some have suggested that the relative resistance of Europeans to HIV, compared with other parts of the world, is because *Yersinia pestis*,[5] the agent of plague, and HIV both attach to cells via the same protein molecule on the target cell surface. People with an altered cell surface protein would be resistant to plague so that their survival compared with that of others would enrich the population with people whose cells could bind neither *Yersinia* nor HIV. The well-chronicled outbreak of the Plague in Eyam in Derbyshire showed that some people had a natural resistance to the disease. Elizabeth Hancock nursed and then buried her husband and all six children, but did not succumb to the disease herself.

Cholera is another toxin disease that changed history. Its sudden arrival in vast epidemics, which afflicted both Europe and the

Americas in the nineteenth century, jolted society into action. People were used to a high, but constant, death rate. Although they did not generally expect to live long, the horrifying and sudden nature of a cholera death induced widespread terror. One government response was to hold a 'National Day of Fasting and Humiliation', during which people were to admit their sins and pray that God would prevent them from getting cholera. This was widely ridiculed. More importantly, a Board of Health was established which produced a novel and ground-breaking report that soon led to improved sanitation and actually did something to tackle the carnage wrought by infectious disease. These first attempts to deal with public health in the UK were taken up on an international scale, and years later would result in the formation of the WHO. The world pandemics that began in the early nineteenth century continue to sweep around the globe. The present one, the seventh, began in 1961 and was a new strain, called El Tor. In 1992, a more dangerous variant of it appeared in Bangladesh, which killed up to 20 per cent of those affected. Apart from the loss of life, the economics of lost production and the effects on tourism are enormous.

For centuries the battle casualties from infectious disease far outstripped those caused by manmade weapons and often affected the outcome of battles. Before the battle of Crécy in the fourteenth century the French referred to the English as the 'bare-bottomed army', because they were constantly squatting to defecate—probably from dysentery or typhoid. In the American Civil War over 60 per cent of the deaths were attributable to dysentery, typhoid, and cholera—all diseases caused by toxins. Similarly, in the Crimean War, over five times more died of such diseases, including cholera, than caused by any weapon. Even in the Boer War at the end of the nineteenth century, when so much had been learned about infection, the percentage of deaths caused by these diseases was the same. Wounds contaminated by soil bacteria, causing gangrene, killed many in the trenches of the First World War. The royal commanders also suffered. Richard the Lion Heart died from an arrow wound that became infected, dying days later from gangrene. The death from dysentery of the Black Prince at the time of the Black Death brought the ineffective Richard II to the throne at the age of 10 and led to the Wars of the Roses which bubbled over for 100 years. Likewise, Henry V died of

What call'd the young, the stout, the gay?
What was it snatch'd the old away?
What made such havoc in a day?
    The Cholera.

What fill'd the people's minds with dread?
What made them sleepless on their bed?
What swell'd the numbers of the dead?
    The Cholera.

What baffl'd all the Doctors' skill?
What did the Hospital so fill?
What was it made our Town so still?
    The Cholera.

Cholera Statement.

Population of Bilston, in 1832 14,492
No. of Persons attacked by Cholera... 3,568
No. of Persons who died by Cholera.... 745
No. of Widowers who lost their Wives
    by Cholera,....................... 103
No. of Widows who lost their Husbands
    by Cholera,....................... 131
No. of Orphans under 12 years of age,.. 450

The first Case August 3rd.
The last Case September 18th, 1832.
Amount of Subscriptions received.. £ 8,536.

Reflections.

Ought we not to view the late fatal disease, which made such havoc amongst us, carrying away our friends and neighbours so rapidly, as a direct visitation of the Almighty? Should we not enquire whether the design of the Lord has been answered? Are we a more obedient and thankful people? Is there less Drunkenness, Sabbath-breaking, and such like disgraceful scenes of dissipation, and works of unrighteousness witnessed in our Streets? Shall we ever again hear of the cruel and inhuman practice of Bull-baiting? Were not amongst the first victims of Cholera, Drunkards, Bull-baiters, and characters of this description? We may forget the calls of the Almighty, we may harden our hearts, and impiously contemn the Lord; but 'woe unto him that striveth with his Maker'! We ought to feel grateful to God, who in the midst of death, preserved us alive; and we are this day 'the living, (may we be) the living to praise him.' During the Cholera, the House of God was considered a sanctuary,—his people the excellent of the earth,—and the means of grace esteemed and frequented. Why should it ever be otherwise? Oh! let each of us ever consider the salvation of our souls, 'the one thing needful.'

What led the people then to pray?
What did they meet for night and day?
What was the Lord to turn away?
    The Cholera.

What, do we see a happier day?
What, have the people ceas'd to pray?
What now the Lord has turn'd away
    The Cholera?

Come, let us all afresh repent,
Come, let us our whole hearts present,
Come, give them all to him who sent
    The Cholera.

BILSTON:

Printed and Sold by Wm. Hackett, Market Place.—Sold also by J. Etheridge, Church Street.

Price One Penny

1.2   Part of a broad sheet published in 1832 with the suggestion that it would serve as a memorial for lost loved ones.

(Held at Wolverhampton Archives and Local Studies, Wolverhampton, UK ref: DX 680/1: www.wolverhampton.gov.uk/archives.)

dysentery two months before he would have been acclaimed King of France. His nine-month-old son was unable to sustain his hold on France during a 20-year war.

The deliberate deployment of biological weapons is often portrayed as a new threat, but they have been used since antiquity—long before they were understood. Indeed the word toxin comes from the Greek for arrow poison, *toxicon pharmakon*. Although *toxicon* referred to the arrow, the phrase got shortened and corrupted over the centuries, so that toxin came to mean a poison. The smearing of putrefying material on arrow tips was a highly effective way to kill people, although it lacked the sophistication of the murderous attack on the unfortunate Georgi Markov.

2

THE GERM OF AN IDEA
A gradual acceleration up to the mid-1850s

Diseases caused by toxins were among the earliest described. Reference to what we now know as cholera is outlined in early Chinese medicine. However, the first detailed accounts of disease that are available today are those of the Greeks, at the time that medicine really began. Diphtheria was described over 2,000 years ago, and the poet Virgil wrote an account of anthrax. Hippocrates (460–379 BC), who is often acclaimed as the 'father of medicine', outlined in detail the symptoms of tetanus[1]:

The master of a large ship crushed the index finger of his right hand with the anchor. Seven days later a somewhat foul discharge appeared; then trouble with his tongue—he complained that he could not speak properly ... his jaws became depressed together, his teeth were locked, then symptoms appeared in his neck; on the third day opisthotonus* appeared with sweating. Six days after the diagnosis was made he died.

The striking clenching of the muscles caused by tetanus greatly helped that particular disease to be defined, but other diseases give vaguer symptoms. For example, we now know that gastrointestinal infections can be caused by many different types of infectious agent. Such illnesses were often either lumped together and described vaguely as 'fevers' or overly subdivided into separate diseases, because of what we now know to be the different reactions of individuals to the same infectious agent. Neither phenomenon was helpful in working out the cause of disease.

A highly significant concept necessary for the understanding of infectious disease is, not surprisingly, that of infection or contagion,

*Extreme arching backwards of the spine and neck.

that is, that disease is caused by some sort of agent that can be caught. The difference between contagion and infection is important. The earliest supporters of the idea that some diseases could be caught thought only of spread from person to person (contagion). On the other hand, infection is a broader term that includes diseases that can be caught from other sources, such as food or drinking water. Before the discovery of small infectious agents such as bacteria and viruses, it was hard to envisage what was meant by contagion. The narrow concept of considering only spread between people would later lead to more confusion and debate.

The theory of infectious disease went in and out of fashion over the centuries. Four thousand years ago, the Mesopotamians had preached cleanliness and used WCs and sewage systems, suggesting that they knew (or guessed correctly) about infection. Galen of Pergamum, a famous and influential Greek doctor of the second century AD, is known to have decamped from Rome to avoid a plague. The devastating Black Death that destroyed much of the European population in 1348 was seen as infectious. The tragic and moving story of the closure of the Derbyshire village of Eyam during the Great Plague, to prevent the spread of the disease, is further witness to that view. Food was left for the villagers outside the parish and paid for in coins that had been dipped in vinegar to disinfect them. Girolamo Fracastoro (Fracastorian), the Italian poet and physician, who among other achievements named syphilis, produced the first scientific theory of infectious disease in 1546. However, this was largely ignored because isolation of infected individuals during some epidemics was not effective in preventing the disease from spreading. Another aspect of the problem was the vague metaphysical suggestions of how disease was caused, for example, Walter Bruel in 1632 ascribed the cause of plague to 'influence of the Starres' and, until the mid-nineteenth century, it was thought that bad smells directly caused disease.

The other critical issue was the debate between those who believed that life could arise spontaneously from other unrelated material and those who believed that all life forms had parents—the germ theory of life. With our modern knowledge of microbes, spontaneous generation may now appear unbelievable, but years ago many people believed that beetles and wasps were generated unprompted from cow dung, or that caterpillars arose from leaves and frogs from mud.

One eminent scientist believed that mice could arise from placing dirty linen and some wheat in a box and then waiting a few days. In contrast to this, in 1668 Francesco Redi demonstrated that, when flies were excluded from rotting meat by fine cloth, maggots did not 'arise', and so he suggested that the appearance of maggots was linked to flies that had landed on the meat.

As is so often the case, a technological advance was to provide the impetus for a scientific one. In this case it was the design of a microscope that was powerful enough and had good enough optics to enable bacteria to be seen. The startling observations made by the Dutchman, Antoni van Leeuwenhoek, from the 1670s onwards marked the beginning of bacteriology. Although often referred to as the 'father of microbiology', van Leeuwenhoek has also sometimes been portrayed as something of a yokel. He was a self-taught shop-keeper, who could speak only Dutch. He ground simple lenses as a hobby and sent quaintly worded descriptions of his observations as letters to the newly formed Royal Society[2] in London and to others. He never wrote a scientific paper. However, he was a model of scientific method in his long and sometimes personal letters, always careful to distinguish fact from supposition. His investigations produced important results in many areas of biology, and his microscopes were of a quality that would not be surpassed for 100 years.

Van Leeuwenhoek was born in 1632 in Delft. At 16 he began an apprenticeship in the Cloth Workers' Guild in Amsterdam, returning when aged 22 to Delft where he bought a house in which he stayed for the rest of his life. He set up as a draper, married, had five children (of whom only one survived him) and advanced as a civil servant in Delft, becoming the equivalent of Inspector of Weights and Measures because of his mathematical skills. After his first wife died in 1666, he married Cornelia Swalmius, an educated woman who may have encouraged his growing scientific interests. A major inspiration for his interest in microscopy is thought to have been Robert Hooke's[3] ground-breaking book *Micrographia* which described microscopical observations, including a description of a microscopic fungus. Another factor that may have affected the time van Leeuwenhoek could devote to his scientific hobby was his relative financial independence by this time, partly from his mother's inheritance and partly from his pay for his civic duties.

These are suppositions. What is indisputable is that from 1673 until hours before his death in 1723 he described his observations in a stream of 190 letters to the Royal Society, following an introduction via his friend the anatomist, Dr Regnier de Graaf. Van Leeuwenhoek was also championed by the poet Constantijn Huygens, father of the physicist Christiaan Huygens, who wrote to Robert Hooke to recommend that van Leeuwenhoek be taken seriously. Hooke served van Leeuwenhoek in another way, in that he was able to reproduce some of the Dutchman's key observations and thus give them credibility.

Microbiology began properly in 1676 with van Leeuwenhoek's identification of 'the smallest sort of animalcules'* in water containing ground pepper[4] in the famous 17-page 'letter 18' to the Royal Society. He subsequently discovered bacteria in his own excreta and that of animals. In 1683 he described bacteria from teeth, finding[5]:

... with great wonder, that in the said matter there were many very little living animalcules, very prettily a-moving ...

He linked these to bad breath, and an absence of cleaning (in those days a toothpick and rubbing with salt and a cloth), leading to a conclusion that pointed forward to future studies of infection:

... there are more animals living in the scum on the teeth in a man's mouth, than there are men in the whole kingdom—especially in those who don't ever clean their teeth, whereby such a stench comes from the mouth of many of 'em, that you can scarce bear to talk to them ...

Van Leeuwenhoek's investigations displayed his flair for experimental design. He realised that the animalcules he observed in rain water could have come from the roof of his house, so he then used a clean bowl placed on a barrel in his courtyard to avoid splashes. The first lot of water was discarded in case the bowl had been contaminated with bacteria. He examined the next batch of water, which showed no sign of living creatures. Not only did this show that the rain did not contain animalcules, but this water could then be mixed with different samples without contaminating them.[6] Experimenting with a vinegar mouth rinse he found that it did not kill all the bacteria stuck to his

---

*Van Leeuwenhoek's word for any of the very small forms of life that he discovered with his lenses.

teeth. This was because the vinegar could not penetrate the layer of plaque to reach all the bacteria. When the bacteria had been scraped off his teeth and were no longer in a thick layer, vinegar was able to kill them. He also sealed up a tube of pepper water and to his surprise still found bacteria. He had discovered anaerobic bacteria that grow only in the absence of oxygen, although he did not understand the significance of this discovery, as it would be 100 years before the discovery of oxygen. Pasteur would 're-discover' these bacteria almost 300 years later. Van Leeuwenhoek has been criticised for not realising the link between the bacteria that he observed and infectious disease. Such suggestions seem unfair and indeed in recent years there has been increasing interest in considering bacteria as part of the normal environment and not just as agents of disease.

Van Leeuwenhoek also made advances in other aspects of biology. Many of these were directly concerned with refuting theories of spontaneous generation. He made detailed observations on the life cycle of the flea, ants and various shellfish, and plant seeds. In addition he examined, with some reluctance, semen from various animals including his own and found millions of animalcules with tails—now known of course to be sperm. He isolated specimens to observe the hatching of flies and dissected testicles to show the origin of sperm. His observations on these larger animalcules were followed up over the subsequent century, but continuing limitations in microscope design hindered an appreciation of the importance of bacteria.

One major key to his success was the exceptional quality of his lenses—it is thought that he made over 500 of them. He was highly secretive about his novel lens design, but on his death donated 26 of his silver microscopes to the Royal Society. Two centuries later the society inexplicably managed to lose them all—a strange episode in its history. It is thought that only around 10 of his original microscopes now survive, several of which are in his native country, the Netherlands.

At first sight van Leeuwenhoek's design of microscope seems primitive, when compared with earlier microscopes that had two or more lenses held in a tube. However, these were optically inferior to his deceptively simple design.[7] Van Leeuwenhoek's advance was to use one very small but high-quality lens. This was held in a small hole between two brass or silver plates, with the specimen held on a needle

2.1 A van Leeuwenhoek microscope, one of the nine remaining microscopes known to be in existence.

(Photograph supplied by the Museum Boerhaave in Leiden, The Netherlands.)

on one side (Figure 2.1). The microscope was difficult to use because it had to be held very close to the eye and adjustment of the specimen was at best crude. However, with a high-quality lens it could resolve objects as small as 1 μm—one-thousandth of a millimetre. This high resolution would have been necessary to be able to see bacteria that are in this size range. He ground some of his lenses, but it is thought that his best lenses, which he was reluctant to let anyone see, were prepared from blown glass using the small, thickened drop of glass that forms by gravity at the bottom of any blown glass bulb. Microscopes today have the magnification neatly marked on the side of the lenses, so that the total magnification and thus the size of the specimen can be calculated. Obviously that was not an option open to van Leeuwenhoek because each lens was an individual one, and had not been produced according to a scientific formula. Again he showed his ingenuity. He estimated the size of his animalcules by comparison to hair and grains of sand.

In 1680 van Leeuwenhoek was elected as a member of the Royal Society, an honour of which he was highly proud. His diploma was in Dutch in deference to his inability to read Latin. His meticulous work had established the existence of small forms of life, setting a firm foundation for the subsequent work that would identify bacteria, and bacterial toxins, with disease. It also put the debate about spontaneous generation on a firmer and more scientific footing, but the battle had not yet been won. However, it was not long before others began to suggest that the small life forms that he had observed could have a role in disease. For example, in 1683 Dr Fred Slare, in discussing an outbreak of animal disease, expressed the view[5]:

I wish Mr Leeuwenhoek had been present at the *dissections* of these infected Animals, I am perswaded he would have discovered some strange *Insect* or other in them.

Others made similar suggestions. In 1720, Dr Benjamin Marten suggested, without any additional evidence, that tuberculosis was caused by bacteria. In the 1760s, Marko Plenčič in Vienna foresaw the specificity of individual diseases, and guessed at the role of infectious agents, although he did not add any new evidence to the debate. His suggestion that:

... nothing else but living organisms can cause disease ...

was too sweeping and speculative a view, and such attitudes led later to discredited attempts to explain all disease as infectious (see Robinson, 1935). The problem was the difficulty in distinguishing different types of bacteria. This led to confusion. Several people tried to separate bacteria on the basis of shape and movement—in particular the Danish naturalist, Otto Friedrich Müller. However, it would be a further 150 years before bacteria would attract serious attention again, when microscopical analysis would link specific bacteria with specific diseases.

Despite the absence of obvious evidence of defined infectious agents, it was still widely held that disease could be infectious. For example, a Dr Mead, writing in 1720, suggested that:

Contagion is propagated by three Causes: *the Air; Diseased Persons*; and *Goods transported from infected Places.*

His views were violently opposed in a pamphlet by George Pye a year later. He did not like such ideas being put about—they might impede commerce and social interactions. He thought the real cause was 'An unhealthy Constitution of the Air, and an unwholesome diet'.

Sir John Colbatch, another London doctor, drew up plans in the same year to fight the recurring threat of plague. This was based on the theory that it was a contagious disease. One part of his plan concerned 'families of substance'[8]:

... but that it shall be Death for any Well Person to come out of such House without a white Wand in his Hand, to warn all People that he belongs to an Infected Family. And those that recover of the Disease, not to associate themselves with other People till after a regular Quarantine ...[9]

Despite the laudable aim of preventing the spread of disease, such constraints were not to apply to the doctors treating the sick, because another section of Sir John's scheme stated 'that no Physician be so far restrain'd to the Limits of his own District'. So much for consistency!

There is one more main character to introduce before we head for the nineteenth century, that of the Italian, Lazzaro Spallanzani (1729–1799). Born just after the death of van Leeuwenhoek, he was ordained a priest, but became a career scientist. He strongly opposed the theory of spontaneous generation. Over several years, Spallanzani conducted a heated debate with another priest—the Englishman John

Turberville Needham. Their experiments concerned infusions—meat or vegetable matter heated in water to provide a rich broth or food that can support the growth of bacteria and other small life forms. These infusions were boiled in flasks to kill anything living and then the flasks were sealed. In Needham's hands such sealed flasks yielded life, and so in his view proved that life could arise spontaneously. Nothing grew in Spallanzani's experiments. In an extended debate, Spallanzani showed that, as the boiled and sealed flasks cooled, the air above the liquid contracted and air was sucked into the flask past the cork seal. This air was the source of the contaminating bacteria that produced the observed growth. There was no need to invoke a 'vegetative force', as claimed by Needham. Despite being championed by Voltaire, Spallanzani's work did not prove to be the breakthrough that won common agreement, and the spontaneous generation argument was not over by the time of his death.

The degree of scientific activity and understanding during the first half of the nineteenth century marked it out from the preceding centuries, although there were still fundamental confusions that had to be resolved. Around 1800 many of those who believed that diseases were contagious did not think that a living entity was involved, because of the widely held view that disease was caused by miasmas or foul-smelling air. Several pieces of work were very suggestive of the central role of bacteria in disease, but did not provide conclusive proof. In 1823, Bortholomeo Bizio investigated the appearance of red stains on moist bread and found that the cause was a small red bacterium that required warmth and moisture. This had previously been given religious significance as the 'blood of Christ'. Twenty years later, Christian Ehrenberg showed that these bacteria could be grown on the surface of potatoes and cheese. Ehrenberg is also thought to have been the first to use the name 'bacterium'. In 1837, Alfred Donné observed bacteria in wounds in syphilis patients and suggested that these might be related to the disease. Work on fungal infections about the same time also suggested a link between microbes and disease. Similarly Hermann Klencke and Jean-Antoine Villemin showed that tuberculosis was contagious, although this work was not widely known about at the time.

In 1847, the quiet and unassuming Hungarian doctor Ignaz Semmelweis was working in Vienna when he made a shrewd observation about the infectious nature of disease. Most importantly, he immediately turned this to practical use. Semmelweis had been appalled by the high level of fatal puerperal[10] fever in women in the hospital where he was a physician. Many women contracted the disease after giving birth. When a close friend died of similar feverish symptoms after cutting his finger while carrying out a postmortem examination, Semmelweis wondered if there was a connection. He noted that doctors moved directly from the postmortem room to the delivery room. He instigated a policy of hand washing and mortality fell from around 15 per cent to 2 per cent. Unfortunately Semmelweis appears not to have promoted his theory very well. As inaccurate reports of his work leaked out, the medical establishment closed in on him, blocking further promotion and pouring scorn on his ideas. In 1865, after he had suffered a breakdown and had been committed to a mental hospital, he himself died at the age of 47 of a fatal infection following a cut finger. Around the same time, Dr Thomas Watson at King's College Hospital in London, and Dr Oliver Wendell Holmes in Boston, each reached the same conclusion that doctors and nurses were spreading the disease. They also suffered ridicule and were ignored. Much later, Lister, whose work followed on from that of Semmelweis, was initially to suffer a similar fate. Even today, the arrogant refusal of some doctors and other hospital staff to follow approved infection control measures is one of the reasons for the epidemic of hospital infections.

It was the analysis of the severe epidemics of the bacterial toxin disease cholera that provided conclusive evidence that some diseases were infectious. The appearance of cholera in the west was sudden. The disease had been around in India for years, but it appears to have afflicted the rest of the world only from the 1800s, causing various pandemics throughout the nineteenth century. Cholera in India is estimated to have killed about 40 million people in the nineteenth century. One suggestion for the appearance of cholera in the west was that it was a punishment from God for those 'whoring and drinking'. A more likely cause for cholera and other infectious diseases was the increased travel and a growing population that was frequently housed in crowded and insanitary conditions. In London at the beginning of the nineteenth century the death rate far outstripped the birth rate and the population increased only because of the continual influx of people to the city. At that time, sewage was dumped in rivers that were also the source of drinking water. Indeed the poor in towns often lived with their own excreta which was stored in houses or gathered in piles in the street—to be shovelled away when the mound grew too high, or the smell too great. Public health was a concept that had not been invented.

The first reported spread of cholera outside India was in 1817. It moved through China, then the Philippines, and soon headed towards Iran and Turkey. It first struck Britain in 1831, then the rest of Europe and the Americas by 1832, where there was a devastating outbreak in New York City. This first epidemic killed about 60,000 in the UK. The nature of the disease induced terror. Its onset was sudden and the patient could be dead in hours. From starting to feel dizzy and cold, the key symptom was sudden and violent diarrhoea and vomiting. The loss of fluid was so massive that visiting doctors would sometimes find the floor of the patient's bedroom awash with a liquid that we now know must have been highly infectious. The draining of liquid from the body made the skin turn blue or black, and the patient look wrinkled. Some exotic remedies were put forward, such as cold water and ice, hot air, bleeding with leeches, black pepper, and ground ginger, but these were universally ineffective. One London dentist even suggested firing cannons every hour to cleanse the atmosphere. A New York physician recommended treating patients with a solution of salts, which is now the preferred method of treatment, although this was not taken up at the time.

The impending onset of the outbreak had prompted the government to set up a Central Board of Health made up of leading doctors. Those initially appointed knew nothing of cholera, but they were later replaced by doctors who had had first-hand experience of the disease abroad. Following this first outbreak, the government was concerned that the rioting seen in the worst affected and poorest areas during the epidemic could lead to wider social disorder. There was also a real fear that the high death rate could lead to a shortage of farm and factory labourers. Concern for health and welfare was not an issue! So a Royal Commission was established to look at the Poor Law that had been set up in the time of Elizabeth I. Edwin Chadwick, a civil servant, was recruited to this commission. He was a barrister, with no training in medicine, who had acted as secretary to the philosopher

Jeremy Bentham, the founder of University College London. Whereas Bentham was a social reformer driven to improve the welfare of the common man, Chadwick's motivation fitted with government thinking. He wanted to cut the cost of supporting families where the man of the house had died of a possibly preventable disease. He was concerned that a workforce that was ill and thus unable to work was inefficient. Regardless of his motivation, Chadwick's contribution to public health was enormous. It was he who ensured that the cause of death was recorded in the new registration of deaths that came into being in 1837. This would ensure that valuable information would be recorded that could later be used to assess the health of the nation. It has been of use ever since.

His groundbreaking and farsighted report was eventually published in 1842 and is generally regarded as the start of public health. It instigated an interest in public health across Europe and the USA. 'The Report of an Enquiry into the Sanitary Conditions of the Labouring Population of Great Britain' was a detailed 457-page document that became an instant hit, selling 10,000 copies. It was read by Queen Victoria at Windsor and was quoted at public meetings across the country. Its statistics showed that labourers and mechanics had a life expectancy of 17 years in Manchester, but 38 in the neighbouring countryside. Indeed 57 per cent of the labouring classes in Manchester died by the age of five. Chadwick added characteristically 'that is, before they can be engaged in factory labour'. Moreover, children left without a working father were thought likely to turn to robbery and prostitution. The report triggered a similar report in the USA to the Massachusetts Sanitary Commission by Lemuel Shattuck, and in Germany influenced Max von Pettenkofer,[11] who designed a sewage system for Munich.

The Chadwick report also dealt with the practical problems of delivering safe water and piping away waste effectively. The engineer Thomas Hawsley introduced the concept of metal pipes, instead of hollowed-out wooden tree trunks, to deliver water under pressure. This was more reliable and ultimately more hygienic. John Roe pioneered the use of the glazed round pipes that are still in use today instead of rough brick channels to remove sewage. This was also much more hygienic. It must be stressed that much of Chadwick's thinking was designed to combat filth and bad smells (thought to be the miasmas), without any thought of infectious disease as we know it. The end result was of course still beneficial.

Chadwick was reportedly a difficult man—he was unsociable and dictatorial. In 1847, he was dismissed as a Poor Law Commissioner, but appointed to an enquiry into sanitation in London. This report led to a controversial Public Health Act. A weakened General Board of Health was resurrected just at the time of the next, and worse, cholera epidemic in 1848–1849. This one killed 130,000 people. By this time Chadwick had upset too many people and was subject to a vigorous attack. He was sacked.[12] *The Times* leader wrote, 'we prefer to take our chance with cholera and the rest than to be bullied into health'.* Many people still subscribe to the view that health, or more properly ill-health, is a personal choice.

Mid-nineteenth century London had expanded rapidly without any proper planning to eliminate human waste and was a smelly unhygienic cesspool. The heavily contaminated Thames, a major source of drinking water, flowed past Parliament choking MPs with its evil smell. Eventually it was decided that London's waste had to be dealt with. The Metropolitan Board of Works, under the control of its brilliant young chief engineer, Sir Joseph Bazalgette, constructed an enormous innovative sewage system that could take London's waste out to the lower reaches of the Thames, where it could be dumped into the tidal river. Bazalgette's far-reaching scheme included the construction of the embankments beneath which the vast main sewers ran and took 18 years to build. It still serves London today.

It was about this time that Dr John Snow suggested, and proved, that cholera was linked to polluted drinking water in his book, *On the Mode of Communication of Cholera* published in 1849. Unfortunately news of his careful efforts did not reach Chadwick, because it would have vindicated his work, even if not his ideas or approach. Snow had become convinced that cholera could be spread by dirty hands and food, and that a further route was via contaminated water supplies. There was a particularly high incidence in one outbreak in London in August 1854, with 344 deaths in four days in the Golden Square area, but hardly any in a neighbouring area. Golden Square is located in London's Soho, just south of Oxford Street. Its water supply was

---

**The Times, 1 August 1854, page 8.

the pump on Broad Street, now named Broadwick Street.[13] Over 500 eventually died in the outbreak. Snow's analysis was supported by his topographical analysis of the deaths (Figure 2.2). He conducted what was the first-ever epidemiological* investigation and found that 87 of the 89 victims whom he examined were known to have drunk water from the Broad Street well. Men working at the brewery in this area, who sensibly drank only beer, did not catch cholera. The local work-house in the area, with its own supply of water, had fewer than expected deaths. It was impossible to reconcile all these facts with the theory of miasmas, because the air smelt bad in surrounding areas and people there did not get cholera. A further particularly strong argument was the case of the Hampstead woman who contracted cholera. She had previously lived in the area served by the Broad Street pump and liked the taste of the water. She regularly had water from the pump collected for her and paid the ultimate price.

Snow's instruction, 'take the handle off the Broad Street pump', was initially opposed as being pointless. However, he prevailed and the handle was removed on 8 September, with the outbreak ending seven days later. It was later shown that a nearby sewer was cracked. Contaminated water from it had seeped into the surface water that was delivered by the pump. Snow's instruction has become famous. Dr David Sacher, the US Surgeon General in 1997, was reported to say 'Where is the handle on this Broad Street pump?' when faced with complex health issues.

The site of the Broad Street pump is now covered by a public house, recently renamed the 'John Snow'—a strange irony, for Snow was a teetotaller for most of his life. The lounge bar of the pub displays some posters explaining the importance of his work and has a visitors' book that has been signed by epidemiologists from around the world. Each September the John Snow Society hold their 'Pump-handle Lecture' and afterwards visit the John Snow pub.

Snow carried out further analysis. He traced the pipelines of water supply companies and showed conclusively the link between supplier and the incidence of cholera. Snow's theories were verified in 1863 during the next and last British cholera epidemic, when unfiltered water was inadvertently released into the public supply, killing 7,000

---

*Epidemiology is the study of disease causation, strictly the study of epidemics.

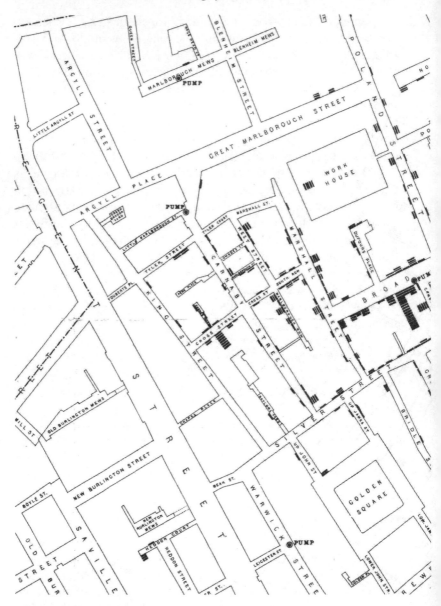

2.2 Part of the map annotated with cholera deaths by John Snow.
The map is available from Ralph Ferichs' website (www.ph.ucla.edu/epi/snow.html).

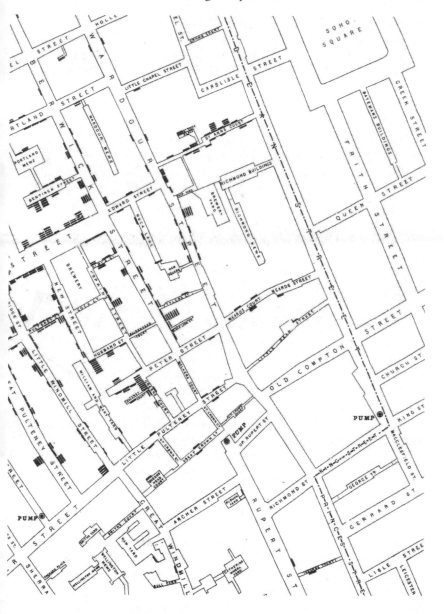

people in the East End of London. John Snow was not alive to see his ideas so clearly confirmed, although his obituary in *The Times* does mention his work on cholera. He died at the age of 45 in 1858 from a stroke. It is speculated that his early death was the result of the other area of expertise for which he was more famous during his life—his pioneering work on anaesthetics. He had administered chloroform to Queen Victoria during the births of two of her children.

Around this time others were carrying out similar analyses.[14] In Oxford, Henry Wentworth Acland, a doctor and Fellow of the Royal Society, analysed an outbreak of cholera there in 1854. He also noted the link of the disease with the water supply. One of the city prisons took its water from close to a sewage outlet. When the supply was closed off, cholera at the prison stopped.

Work on Public Health was continued by John Simon, who had been elected as a Fellow of the Royal Society at the age of 29 for his pioneering work on the thyroid gland. Simon, later Sir John Simon, conducted a similar analysis to that of John Snow. This confirmed Snow's findings and showed that water that had been sand filtered was the safest. His team's investigations were wide ranging, from malnutrition to an early analysis of occupational health. Simon was obviously more politically astute than Chadwick and the work of his team led eventually to the 'Great Public Health Act' of 1875. The cholera epidemics had driven Britain into the vanguard of the new hygiene movement.

The public health and hygiene reforms had made a huge impact on health, but still did not explain how disease arose, or whether bacteria might be involved. Early studies had discovered a 'comma'-shaped bacterium in diarrhoea from cholera patients. In 1854 Filippo Pacini named this bacterium *Vibrio cholerae*. However, bacteria were not conclusively shown to be the cause of the disease until Koch's experiments several years later.

An important unresolved question was whether all bacteria were really the same. If they were, it would be hard to explain how different diseases could be linked to bacteria. Despite van Leeuwenhoek's descriptions of different types of bacteria and the improved microscopes of the nineteenth century, there were two big difficulties. The first came from the observation that fungi could display different forms, and this was assumed by some to apply also to bacteria. Thus the different shapes of bacteria were suggested to be different forms of the same bacterium. This view was put forward particularly strongly by Jean Hallier in Jena, Theodor Billroth in Vienna, and the famous botanist Carl von Nägeli in Munich. Against these unfortunate distractions were ranged the careful work of a number of meticulous scientists. Carl Gustav Ehrenberg published a massive volume differentiating different bacteria on the basis of their shape and ability to move. However, none of his descriptions would be recognised today. The Frenchman Dujardin simplified this scheme, but Ferdinand Cohn, whom we shall meet again in Chapter 3, carried out more important and useful work. He separated out true bacteria from other small organisms. This work provided a solid foundation for future classifications.

The second problem was the small size of bacteria, which made it hard to distinguish different types. Bacteria are so small that it was, and remains, difficult to find many distinguishing features down a microscope. Other techniques were needed and these came later.

At the same time, there was further progress in the battle to defeat spontaneous generation. Following on from Spallanzani at the end of the eighteenth century, Franz Schulze also experimented with boiled extracts that were sealed to prevent contamination by air. He showed that life did not arise if these extracts were supplied with air that had passed through strong acid. Theodor Schwann used a similar approach. He sterilised the air by heating it. Later Georg Schroeder separated his infusions from the contamination in air by using a cotton-wool plug—a technique that is still used today in microbiology laboratories. Joseph Leidy in Philadelphia was another who opposed spontaneous generation. But, although the evidence against spontaneous generation kept accumulating, the controversy could not be resolved.

So, by about 1850, there were many who had the correct ideas about infectious disease, but there was still insufficient proof to link specific bacteria with particular diseases. The gradual accumulation of knowledge over the centuries was about to be replaced by an explosion of activity, leading to a profound understanding of infectious disease. That explosion would be triggered by Louis Pasteur and Robert Koch.

3

THE GOLDEN AGE OF MICROBIOLOGY
Pasteur, Koch, and the birth of the toxin concept

The last 25 years of the nineteenth century have been referred to as the 'golden age of microbiology'. This is no exaggeration. The advances in microbiology at that time exceeded even the great leap in our understanding of bacteria that occurred during the late twentieth century. Infectious disease, not even an accepted concept in 1850, moved from dark religious superstitions to a science. In the process medical microbiology was invented and toxins were discovered. Indeed the latter half of the nineteenth century was a wonderful time for science in general. Charles Darwin discovered evolution by natural selection, while Gregor Mendel laid the foundations of genetics—each impinged on and was relevant to understanding bacteria. Enlightenment in chemistry was perhaps slightly ahead of biology, because modern chemistry was being established by the start of the nineteenth century. However, at the end of the nineteenth century the insights of Mendeleev had put the elements in a logical order called the periodic table and had banished the equally dark fantasies of the alchemists. Physics was also transformed during this time by such as James Clerk Maxwell, Marie Curie, and Max Planck.

The revolution in microbiology in the second half of the nineteenth century is primarily linked to two men: Louis Pasteur and Robert Koch. Louis Pasteur was born in December 1822, the only son of a former soldier. His father had received the cross of the Légion d'Honneur from Napoleon and had returned to Dôle, a village in eastern France close to Switzerland. Pasteur senior was a tanner whose family could be traced back over many generations in the same district. His mother, whose family had long worked on the land, was similarly poor. Pasteur had three sisters whom he seems to have regarded with great affection. This early life appeared to instil in him a devotion to France and to his family that never left him. As a young boy he did not appear particularly quick, but his headmaster noted his diligence and striking attention to detail. Encouraged by his headmaster and by his parents' desire that their son should proceed with his education, the 16 year old was sent to Paris. However, he was soon homesick and returned to Dôle. He then attended a nearby college at Besançon where his work was good without any display of brilliance. At his second attempt he gained entry in 1843 to the École Normale Supérieure, the famous Parisian college founded during the French Revolution, to study chemistry. He was a serious student,[1] fond of reading and talking of crystals and mathematics, and reading philosophy. Indeed he showed a real talent as a painter, as some of the pictures on the walls of the Pasteur Museum demonstrate.

At the École Normale, Pasteur's progress was creditable, without being viewed as outstanding. By 1847 he had obtained his doctorate degree with dissertations in chemistry and physics—much to the pride of his parents. By this time he had exceeded the ambitions that they had for their only son. His first research project after this looked at the peculiar ability of some chemicals to rotate the plane of polarised light.[2] It had been known for some time that solutions of chemicals called tartrates rotated light to the right, whereas others, the so-called *para*tartrates, had no such effect, although they appeared to be chemically identical. Pasteur made crystals of the *para*tartrates and found that there were two characteristic shapes—one was a mirror image of the other. When the different crystals were separated and dissolved in water, one rotated light to the left and one to the right. This was demonstrated to the celebrated physicist and astronomer Jean-Baptiste Biot, who thereafter was one of Pasteur's most ardent champions. Most importantly Pasteur suggested correctly that this asymmetry might reflect the arrangement of the atoms in the compound. This work opened up a new area of chemistry, stereochemistry, for which he was awarded a prize of 1,500 francs by the Paris Pharmaceutical Society and received the Red Ribbon of the Légion d'Honneur.

His next experiment was a biological one. It was known that contamination by biological matter could cause tartrates to be fermented to produce other compounds. Pasteur showed that his optically inert mixture of right- and left-rotating compounds could also be fermented, although only the right-rotating tartrate was worked on, leaving the left-rotating tartrate untouched by the swarming micro-organisms. We now know that this is because he was correct in his view that the light-rotatory properties of these chemicals did reflect the arrangement of the atoms. Biological processes, such as fermentation, work because proteins called enzymes bind to compounds to facilitate their chemical reactions.[3] The interaction between enzyme and chemical is so specific that an enzyme can distinguish between the different arrangement of atoms in the right- and left-rotating compounds. This was the point where Pasteur crossed the boundary between chemistry and biology and was then able to apply his trained chemist's mind to attacking biological problems.

Meanwhile Pasteur had moved to Strasbourg to become Professor of Chemistry, although he had to use some of his prize to furnish his new laboratory. He was very busy and wrote home that he would not marry for a long time. This careful and perhaps naïve planning was disturbed by Marie Laurent, the daughter of the Rector of the Academy, and within two weeks of meeting her Pasteur was proposing marriage. This hasty behaviour was perhaps further proof of his supreme self-confidence. They were married within months and throughout their long marriage Madame Pasteur appeared willing to play a supportive role as her husband's fame spread.

Within a short time Pasteur moved to Lille as Dean of the Faculty of Science, where he appears to have been a successful and innovative administrator. In his opening address as Dean he quoted Benjamin Franklin's reply when challenged about the value of basic science: 'What is the use of a baby?' The answer of course is that it has little immediate use, but that it inspires hope and has enormous potential. Pasteur was not just a meticulous experimentalist, but a philosopher of science who used his rhetorical skills to persuade others and to further his own viewpoint. Throughout his career Pasteur always championed basic science for its value to humanity, while at the same time often becoming involved in research of the greatest practical significance. He also produced his own oft-repeated exposition on the fickle and unexpected nature of science: 'In the field of observation, chance favours only the mind which is prepared.' Certainly, many of Pasteur's discoveries apparently arose from unexpected quarters, as commonly happens in science.

Now that his attention had been caught by bacteria, he wanted to know more about them. The prevailing view was that the rotting and putrefaction that turned meat bad was caused by the death of the organisms that were present, and was not related to their life. Similar views were held about the fermentations that changed sugar into the alcohol in beer and wine. In the former case, this was partly because yeasts were not always found—people had overlooked the much smaller bacteria that Pasteur showed were present in rotting meat. He showed that all such putrefactions and fermentations, as in beer production, were linked to the presence of living organisms. They had nothing to do with dead ones. It was simply that different bacteria or yeasts produced different compounds. The only difference between fermentations and putrefactions was that the former made compounds that were useful such as beer and wine, whereas the latter made products that were not wanted and were very obviously smelly.

During this work Pasteur made another major discovery. Bacteria squashed between two pieces of glass so that they can be looked at in the microscope are usually most lively at the edge where there is a plentiful supply of oxygen from the air. Pasteur found that some types of bacteria were most lively at the centre where there was least oxygen. He had discovered anaerobic bacteria, originally seen by van Leeuwenhoek almost 300 years earlier. Whereas science had not been sufficiently advanced for van Leeuwenhoek to appreciate what he was observing, Pasteur was able to understand the significance of bacteria that grow only where there is little or no oxygen.[4]

Pasteur also showed that a solution of the correct sort of inorganic (that is, non-living) salts was sufficient for yeast to change sugar into alcohol. There was no need to postulate that the growth medium contained some life force. Living organisms could perform chemistry—it was as uncomplicated as that. His fermentation experiments brought him directly to consider the ever-present theory of spontaneous generation, although his friend and mentor Biot advised him against entering this highly contentious issue. Pasteur by this time had returned to the École Normale Supérieure as Administrator and Director of Scientific Studies,[5] and he was not to be dissuaded.

At every twist and turn of the argument, the spontaneous generationalists found some new excuse to support their view. The latest proponent was the celebrated naturalist Félix Archimède Pouchet, who was at the time Director of the Natural History Museum at Rouen. In 1859, he criticised the earlier work on the boiled extracts of Schulz and Schwann, suggesting with apparently carefully controlled experiments that life could arise spontaneously. Pasteur set out to refute this in as definitive a manner as possible, producing over the next few years a series of papers that attempted to settle the matter once and for all. He extended the work of early nineteenth-century scientists by using flasks with very long drawn-out necks, called swan necks (Figure 3.1) for obvious reasons. As had been done throughout this old controversy, the rich infusions in the flask were boiled, but in his experiments the flasks were not sealed. As they cooled, the air in the flask contracted, slowly drawing air into the flask from the outside through the long swan neck. Pasteur reasoned that any bacteria in the air would stick to the walls of the narrow neck and not reach the infusions, which would remain sterile. This is exactly what happened. Furthermore, bacterial growth was observed when the flask was later tilted so that liquid washed up the tube. The advantage of this experiment over earlier ones was that the spontaneous generationalists could not argue that either heating or passage through strong acid had destroyed some life-giving vital force in the air. Later Pasteur showed that body liquids such as blood and urine could be kept in flasks without any growth of organisms if they were carefully withdrawn from the animals to avoid contamination. As this totally avoided the use of heat it proved that the postulated 'life force' inherent in materials that led to spontaneous generation did not exist.

Pasteur was the first to look for and show the presence of bacteria in the air. This had been assumed, but not proved, by others during the spontaneous generation debate. He tested the air in different places for the presence of bacteria. On his holiday in the Alps he found that the air contained few bacteria compared with the yard outside his laboratory. Pouchet answered these experiments with Alpine tests of his own and obtained different results. However, Pouchet used boiled infusions of hay—this was an unfortunate choice because much later it was shown that the bacteria in hay were resistant to boiling.

A special Commission of the Academy of Sciences was set up to investigate who was correct. Pouchet correctly judged that the Commission were highly biased against him. He refused to comply with the conditions, leaving Pasteur to win rather unconvincingly on this

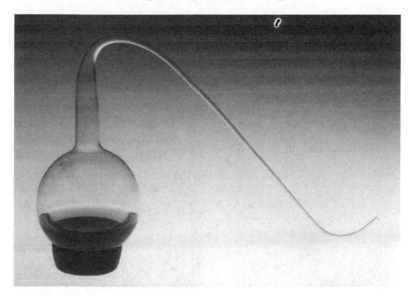

3.1 A swan-neck flask used by Louis Pasteur in his experiments to prove the germ theory of life.

(Photograph supplied by the Pasteur Institute in Paris, France. © Institut Pasteur.)

technicality. This was not at all satisfactory. If put under pressure to repeat Pouchet's experiment, Pasteur would presumably have found similar results and would have had to investigate the hay infusions further. The problem was properly solved 14 years later after the Englishman Charlton Bastian suggested that boiled urine would support the spontaneous generation of life. Pasteur was provoked to refute this and, after much effort, he and his colleagues found an explanation that applied to both Pouchet's and Bastian's experiments: some bacteria were very resistant to heat, and could not be killed by heating to 100°C. Neither the boiled urine in Bastian's experiments nor the boiled infusions of Pouchet were sterile. Sterility can be guaranteed only by heating to a higher temperature under pressure.

Pasteur's view was also strongly supported by some clever clearcut experiments by the English physicist John Tyndall. In addition, Ferdinand Cohn showed that some bacteria made spores that were very heat resistant. From this point on, interest in spontaneous generation dwindled, partly because of the compelling nature of the

3.2 Louis Pasteur and Robert Koch: Pasteur, aged 58, photographed in London 1881. (© Institute Pasteur.) Koch, aged 54, photographed in India 1897. (This photograph was given to the author when he visited the Indian Veterinary Institute in 1988.)

experiments that refuted it. These were discussed widely in the press. In addition, as specific bacteria became associated with specific diseases in the 1870s, the spontaneous generation theory became less tenable. The theory, fantastic although it might seem now, had survived for centuries attracting widespread general and religious interest, but was finally defeated. Pasteur had played a key part in its defeat for several reasons. First, the comprehensive nature of his attack on the problem covered several angles at once. Second, he forcibly defended his position against any attack. Finally, he played to the public interest very effectively by his eloquent public lectures. However, Pasteur was not entirely open and honest about his own research in this area—choosing to regard his own experiments where life mysteriously appeared as 'unsuccessful' and the result of contamination.

In the next phase of his career, Pasteur tackled four applied problems of great economic importance to France, all of which involved infections of one sort or another—the so-called diseases of vinegar, wine, silkworms, and beer. His help was requested by none other than the Emperor. In each case Pasteur not only found the infectious cause of the problem, but also devised a way to treat it. The solution to the problems with wine and vinegar that went off was to heat them enough to kill off contaminating bacteria that had remained after fermentation. The silkworm disease turned out to be two diseases, which complicated matters for a while. However, Pasteur devised tests to ensure the selection of healthy eggs and so saved the silkworm industry from ruin. The beer industry's difficulties were caused by yeasts that were contaminated by other micro-organisms, and Pasteur was particularly pleased that he could help French beer to compete with the renowned German beers—particularly as the two countries were at war at the time. All these successes further increased his fame.

During this time Pasteur suffered a number of tragedies. Having lost his oldest daughter Jeanne to typhoid in 1859 when she was nine, his own father died in 1865. In the same year Camille, one of his remaining daughters, died aged two, to be followed the next year by Cécile aged 13—again tragically of typhoid.* Around this time Pasteur triggered an explosive confrontation with some of the students at the École Normale, and obstinately refused to retreat from his position, stating that he would resign and close the school. The student body did not give ground; the school was closed and reopened later in 1867 without Pasteur as Director. However, something had to be offered to the now famous Pasteur. He was given a professorship at the Sorbonne and the French government, with the Emperor's backing, agreed to the construction of new laboratories for him at the École Normale for work on infectious diseases. Then, in 1868 at the age of 46 he suffered a stroke and thought he was going to die:

I am sorry to die; I wanted to do much more for my country

His slow convalescence was not helped when he learned that the building work on his new laboratory had been halted on the day he had become ill, because it was thought that he would not recover. The work was resumed at the personal intervention of the Emperor.

---

*It is mere speculation, but the death of two of his children to typhoid may well have further strengthened Pasteur's determination to tackle infectious disease.

Pasteur made a slow recovery, although his health was never the same again. It is remarkable that his most famous discoveries lay in the years ahead.

The defeat of spontaneous generation made it likely that the germ theory could also be true for human disease. The British surgeon Joseph Lister had taken the next step in applying these ideas, by advocating the use of carbolic acid as an antiseptic during surgery to kill off germs that caused postoperative infections. He was well aware of Semmelweis's work on infections in childbirth, and had been impressed by Pasteur's findings. Lister showed that death caused by gangrene and blood poisoning could be reduced by treating wounds and spraying operating theatres with carbolic acid.

After recovering from a nervous breakdown aged 20, Lister had built an impressive career. He was a dedicated and serious man, charming but lacking a sense of humour. As an Englishman, he experienced some resentment, particularly from the Glasgow establishment when he was Professor of Surgery there. He subsequently moved to Edinburgh and then to London at King's College Hospital. He treated Queen Victoria in 1871 and was later elevated to the peerage. Lister was elected a Fellow of the Royal Society and was its President for five years from 1895, being one of the first to achieve the high honour of the Order of Merit. However, this outward appearance of being a glittering part of the establishment is deceptive, because he had to fight for his ideas to be accepted. He was particularly subjected to ridicule in Britain, although latterly his ideas were acknowledged. Carbolic acid does not create a particularly pleasant environment, and by the end of the nineteenth century aseptic techniques were being employed where pathogens are excluded by the use of masks, gloves, and gowns, as opposed to killing what was there.

None of these advances really proved that bacteria caused disease. In 1850 Casamir-Joseph Davaine and Peirre Rayer had found a large bacterium in the blood of cows dying of anthrax. The bacterium was also found in cases of the disease by Aloys Pollender. This observation did not prove that the bacterium caused anthrax, but Davaine later transmitted anthrax to animals by injecting blood containing the bacterium. This was still insufficient proof to convince everyone, especially those who could not repeat the observations. Pasteur entered the fray in support of Davaine, demonstrating that blood containing anthrax could become contaminated, unless carefully collected, leading to confusing results.

Anthrax* is an animal disease that had been known about since ancient times. In the 1600s, the 'Black Blane' killed around 60,000 cattle in Europe. Humans can catch the infection, as skin infections or more dangerously gastrointestinal and inhalational disease, from animals. There is no evidence of human-to-human spread, and in that sense it would not have been thought to be infectious. Those working with animal skins are most at risk. These include tanners exposed to contaminated leather and people sorting wool for the carpet trade who are prone to inhalational anthrax, which is usually fatal and known as woolsorters' disease. In the early twentieth century a health scare with shaving brushes led to several deaths.[6] It was only in the mid-eighteenth century that the disease symptoms were properly described in animals and humans, although at the time a connection was not made between the human and animal diseases.

Although Davaine and Pasteur had made some progress, there were still difficulties that could not be explained. The bacteria disappeared from dead animals, but it was known that dried blood was still infectious. A further complication was that fields where anthrax-infected animals had grazed appeared to harbour the ability to cause the disease for many years.

Meanwhile to the east, in a small town called Wollstein that is now located in Poland, a new star was about to arise. A country doctor called Heinrich Herrmann Robert Koch began studying anthrax in a curtained-off area of his consulting room.

Robert Koch was one of a family of 13. He was born in 1843 in the city of Claustal, where his father was head of the local mine. Robert was reportedly his mother's favourite son. However, it appears that Koch's Uncle Eduard, a keen naturalist and a very early enthusiast for photography, was an important influence. Koch was diligent at school, although, like Pasteur, not noted as outstanding. Originally intending to become a teacher, Koch switched to medicine, and by 1866 he was qualified, engaged to Emmy Fratz, and thinking of settling down as a family doctor—although even then he really wanted to travel like two

---

*The word anthrax is from the Greek for coal because of the black skin lesion that is produced—a painless ulcer about an inch across with a black centre.

of his brothers. His early years in practice were fraught with difficulties, but by 1869 he was building up a practice in Rakwitz (now part of Poland), and briefly served as a doctor in the Franco-German war of 1870. There was no obvious suggestion of his future directions.

In 1872 he became the District Medical Officer in Wollstein. This entailed some extra duties as well as his own medical practice. It is clear that the young Robert Koch enjoyed his time there: a popular doctor, respected in the community, adoring his growing daughter Gertrud, and even having time for some amateur archaeology. Pictures of him suggest that he was stern and serious,* although he has been reported to have had a good sense of humour, and was a bit of a *bon viveur*, witnessed by his beer drinking with friends and his love of good food and wine. But, most importantly, his extra income enabled him to buy a microscope† and pursue his interests in natural history. He then began his part-time studies in what was to become medical microbiology. Initially his work on anthrax was carried out in the same room that he used to see his patients. Later his wife built a curtain to divide the room into two, so that his patients could at least enjoy some semblance of separation from his cultures of anthrax and his experimental animals. These animals were housed in the garden and looked after by his wife and daughter.

A few years after settling in Wollstein, Koch embarked on an extended jaunt visiting several conferences and laboratories, as well as museums, restaurants and beer houses. He returned invigorated. He was ready to tackle anew a problem that he had begun to work on before leaving—the outbreak of anthrax that was affecting sheep and cattle locally and that also led to anthrax in humans. Koch observed that the bacteria appeared to change their optical properties. He was observing the signs of spore formation—the ability of the anthrax bacterium to change into a heat-resistant dormant state. In a period of about a month Koch conclusively worked out the life cycle of anthrax. In addition he set in train a novel and revolutionary experimental approach. He noted that the bacteria grew well in the eye fluid of an infected rabbit, so he obtained cattle eyes from the local abattoir

---

*In his latter years Koch has been described as aloof and arrogant by some, but not all, contemporary witnesses.

†Koch was one of the first to use the Abbé condenser, which represented a great advance in microscope design.

for the artificial culture of the bacteria. Having observed spore formation under these conditions, he correctly inferred that this could explain many aspects of the disease—the longevity of infection in the soil and the danger of handling animals that had died of anthrax. Even when there were no living bacterial rods, the spores could survive for a long time and then spread the disease. It is interesting that Koch referred to the bacterium by the name *Bacillus anthracis*, indicating the linkage of a specific disease to a particular bacterium. Koch realised at once the implications of his work for public health.

However, he was unsure whether his conclusions were correct—after all he was a country doctor working in isolation from the scientific community. Koch wrote to Ferdinand Cohn in nearby Breslau, seeking support for his conclusions. Cohn, who had done so much to classify bacteria, invited him to demonstrate his experiments, and was immediately won over to become an enthusiastic supporter of Koch's work. Cohn had also observed spores, although in the harmless but similar bacterium, *Bacillus subtilis*. Two papers describing Koch's and Cohn's work were published back to back in Cohn's journal in 1876.

Koch now turned to improving and developing staining techniques for bacteria that enabled them to be more easily examined in the microscope. In addition, he applied his childhood interest in photography to produce the first-ever photographs of bacteria. These were of outstanding quality, all the more amazing considering the technical difficulties of nineteenth-century photography.

In the meantime, work on anthrax passed over to Louis Pasteur, who was recovering from his stroke. Detractors of Koch's experiments had argued that the disease that he had transmitted to mice with the anthrax spores could have been caused by some contaminant carried over with the bacteria that he had obtained from a diseased animal. Pasteur showed that this was impossible, by growing the anthrax bacteria in urine and inoculating a drop of this into fresh urine. When repeated several times over, the bacteria obtained were still able to cause disease, but the carry-over from the original sample was infinitesimally small. This strongly supported the germ theory of disease. It also illustrated how well Koch and Pasteur complemented each other, although their personal relationship was later so poor.

Pasteur was soon called upon to take his discoveries into the field—in this case literally. The French government requested that he look into outbreaks of 'spontaneous' anthrax, where animals died of anthrax although there was no known contact with infected animals. Once again he solved the problem, and at the same time came up with a practical solution. Animals killed by anthrax were being buried in fields grazed by other animals. Pasteur showed that the little piles of earth dug up by earthworms contained anthrax spores. It was possible to avoid passing on the disease by choosing burial sites for dead animals more carefully.

Pasteur now made a further bold stride into the unknown, the deliberate creation of a vaccine. The popular view of this story is that this approach came from a chance experiment with a different disease, fowl cholera.* His work on this had led to him making a vaccine against this disease. Old cultures of the fowl cholera bacterium, now called *Pasteurella multocida* (the 'multocida' means killing many), had lain around in Pasteur's laboratory and had lost the ability to infect chickens. He decided to reuse the birds that had failed to become ill. When he tried to infect them with freshly grown bacteria, the birds appeared to be immune to infection. He had discovered a way to attenuate, or weaken, bacteria so that they could stimulate the body to recognise the infectious bacteria without causing damage and thus be prepared for the real infection. Once again, he could declare that 'chance favoured the prepared mind'.

Pasteur's announcement of his chicken cholera vaccine did not disclose how it had been weakened (prolonged exposure to air) until nine months later in October 1880. The story then continues that Pasteur applied similar principles to anthrax to produce bacteria with weakened virulence. These bacteria then induced only mild disease and importantly gave protection against infection with fully virulent bacteria. This announcement was made in early 1881, just six months after the work on chicken cholera. It is now clear that Pasteur's declaration of this success was woefully premature and that he was irresponsibly driven to this by the publication of work on anthrax by a now long-forgotten veterinarian, Jean-Joseph Henri Toussaint.

Pasteur's detractors remained unconvinced by his hastily prepared publication on the subject. He was challenged to carry out the same experiment with farm animals by Hippolyte Rossignol, a veterinarian

*Not related to human cholera which is caused by *Vibrio cholerae*.

who was opposed to the germ theory of disease. Pasteur was under pressure to accept. This field trial was conducted at Rossignol's farm at Pouilly-le-Fort, a short distance from Paris. It was a very public experiment, attended at both the start and finish by a crowd of scientists, doctors, farmers, and the press. Twenty-five sheep and six cows were given weakened bacteria as a first vaccine; similar numbers were left untreated as controls. Twelve days later a second vaccine, more virulent than the first, was given. Two weeks later all were given anthrax. Pasteur confidently predicted that all vaccinated animals would survive and all the non-vaccinated ones would die. Two days after this, as a crowd of more than 200 gathered to witness the outcome, Pasteur arrived to be welcomed by applause, as the results were by then obvious. All vaccinated animals survived, while all the non-vaccinated animals were dead or dying. It was beautifully clear cut. Pasteur was to remark that he would have been inconsolable if the discovery had not been a French one! Of course this work was still subject to challenge. One failure was in Italy, although the disbelievers there did not respond to Pasteur's willingness to conduct a local trial. More notable was the German challenge led by Koch, who was keen to give priority to Toussaint's work out of his ill feeling towards Pasteur. One of Pasteur's assistants was sent to Germany and, after an initial failure, the vaccine was shown to be effective and was widely adopted in Germany too.

It now emerges that the reality is not quite as told or implied by Pasteur. He clearly let people believe that the anthrax vaccine had been prepared by his air-attenuation technique, whereas it was prepared by a method developed by his closest colleague, Charles Chamberland. He, along with Émile Roux, had been using chemicals to attenuate bacteria. At the time that Pasteur published that he had a vaccine for anthrax, his own results with air attenuation were giving very variable results—sometimes the vaccine protected the animals, sometimes it killed them. On the other hand, Chamberland's treatment of anthrax with potassium bichromate was more reliable. So when Pasteur rashly accepted Rossignol's challenge he decided to use the chemically treated vaccine. In fact, Pasteur's work on an air-attenuated anthrax vaccine came to fruition during the trial and he conducted a successful trial using it later that summer.

Does this deception matter, given that the eventual outcome was so successful? First, it shows a certain ruthlessness in the way Pasteur treated Toussaint and, second, it is an illustration of Pasteur's prejudice and stubbornness. Toussaint's attempts at an anthrax vaccine were only partially successful, in that some of his vaccinated animals had died. Pasteur's anthrax experiment had established the principle of immunity, but he did not understand the basis of immunity—no one did at that time. Pasteur believed that bacteria would have to be alive to lead to immunity[7] and perhaps use up some chemical that was needed for the pathogen. That Toussaint's vaccine should have had any effect was disturbing for Pasteur, because his vaccine had been prepared by heating and thus was expected to be made only of dead bacteria. Although Pasteur was publicly pointing out the problems of an inactivated approach like that of Toussaint, at the same time he had secretly used one vaccine which, although live, had been chemically treated, and he was also secretly investigating chemical treatments in his own laboratory. Although Pasteur did concede that Toussaint had a role in the anthrax discovery (largely ignored nowadays), he can certainly be accused of sharp practice in his treatment of the more junior man, who became mentally ill a year after the great Pouilly-le-Fort experiment and died several years later. Against that it is clear that Pasteur's work (and that of his colleagues working in the background) was more comprehensive and put across more convincingly with Pasteur's acclaimed verve and determination.

William Greenfield, working at the Brown Animal Sanatory Institution in England, also carried out animal challenge experiments with an attenuated anthrax vaccine. Indeed his experiments were concluded a few months before Pasteur's, but Greenfield's work is also largely forgotten.

Regardless of these considerations, this work was a great triumph for science. In a mere seven years since Koch's classic and definitive work on anthrax, Pasteur had used this knowledge to provide a means to prevent the disease. In addition, this story had provided the first clear link of a bacterium to a disease, and initiated the concept of attenuation that was to be the hallmark of future vaccines. Before this, the only vaccine was cowpox which was used against smallpox. These bacterial vaccines could be described as the first genetically engineered vaccines.[8] By 1883 half a million animals had been vaccinated and protected against anthrax.

The anthrax story drew attention to a conundrum that could not be solved with the knowledge of the time. Koch's linkage of a specific bacterium with anthrax was based on the premise that bacterial species were distinct and stable, and that individual bacterial species could be discriminated from each other by techniques of staining, shape, and growth requirements. This viewpoint had been hard won after much argument by many protagonists over the years. It was supported vigorously and rather rigidly by the 'German school', led by Cohn and Koch. Against this were the 'unitarians' such as Nägeli and Antoine Béchamps, a former assistant of Pasteur, who thought that there were only a few different types of bacteria (that arose spontaneously) and that these could change their shape and characteristics. Their rationale came from the correct observation that some fungi change shape.

The attenuation of bacteria—that is, a change in properties—described by Pasteur and his colleagues did not appear to fit with the concept of specific bacteria with invariant properties. However, the apparent conundrum can now be easily explained. Changes (mutations) in the genes of any living organism can occur at a very low frequency—either because of DNA damage or during copying of the DNA when errors can occur. The environment can select favourable mutations. This is natural selection and is the basis of Darwinian evolution. In the case of Pasteur's attenuated bacteria, prolonged exposure to air or chemical treatment would have selected bacteria able to cope with this stress. These bacteria had coincidentally lost virulence characteristics. These are small differences compared with the considerable differences between different types of bacteria. These details were not known at the time and thus the question of bacterial stability led to further confusion. Unfortunately, the rigid view of Koch, with regard to bacterial stability, was twisted by some to support the extreme view that there were only a few different types of bacteria.[9]

The anthrax work also sparked the beginning of the hatred between Pasteur and Koch. These disagreements were expressed most vigorously—both verbally and in writing. Koch in particular used inflammatory language that did him little credit. He once remarked that 'Pasteur is not even a doctor'. The two men first met in 1881 at a meeting organised by Lister in London, where Koch showed his agar plate technique for growing bacteria. He had established this with the (unacknowledged) help of Fanny Hesse, the wife of one of his colleagues—it was she who had suggested the use of agar.[10] A bacterial sample is smeared across the surface of jelly-like agar containing nutrients. After the plate is kept warm for about a day, the growth and multiplication of bacteria produce a colony of around a thousand million bacteria which can be seen with the naked eye. The crucially important aspect of this is that the colony has arisen from the growth of a single bacterium, and it is an excellent way of separating individual bacteria from any contaminants. This technique revolutionised microbiology and remains indispensable (and unchanged) today. Koch criticised Pasteur's use of liquid cultures, while in return Pasteur pigheadedly criticised the agar technique. And so it went on.

At least part of their antagonism may have stemmed from problems with language. At one meeting, Koch mistakenly thought that Pasteur had referred to one of his papers with the words 'German pride' when in fact he had referred to a 'German collection'. Another problem may have been the striking differences in their background, education, and basic characters: one patriotically French, the other Prussian; one interested in vaccines, the other in hygiene.* It is sad that these two towering figures of science should have clashed in this manner, although they were not the first scientists to express personal and vindictive antagonism. It remains a feature of modern science. Indeed the tension generated by this animosity may have spurred each group to greater efforts to beat their rival and thus advance the cause of science, but it also served to inhibit potentially useful collaboration. Unfortunately this antagonism spread to others in the laboratories of the two protagonists, although it lessened after Pasteur's death and when Koch was no longer active in science. The interaction between the Pastorian and Kochian camps also improved partly as a result of Ilya Metchnikoff's moderating influence.[11] He worked at the Institut Pasteur, although he was keen to break down the antagonisms between the Paris and Berlin schools. Indeed later there was active interaction between scientists from the two laboratories.

Pasteur, now a national hero, carried out some work on the bacteria now known to be *Streptococcus* species—identifying them as the

---

*Departments of Hygiene are still found in Germany, but not really elsewhere, partly as a result of Koch's influence.

cause of boils—before turning to his last and best-known triumph—
that of the vaccine against rabies—a viral disease.[12] The spectacular
and sensational treatment of people bitten abroad who travelled to
Paris in hope of being saved from this dread disease confirmed his
fame. Moreover, grateful donations arrived from across the world
and soon over 2 million francs were available for the construction of a
new Institute—the Institut Pasteur, which opened in 1888. Pasteur,
its first Director until his death, concentrated on the administration
of the great new institute which still proudly bears his name, leaving
his protégés to take forward the science that he had begun.

Pasteur was granted many honours. He was elected to the Academy
of Science in 1862 and even the Academy of Medicine in 1873. This
last office was resented by medical men who were against the idea of
a scientist joining their academy. He was awarded the Grand Cross
of the Légion d'Honneur in 1881. He also received many foreign
honours, such as election as a foreign member of the Royal Society
in 1869. Then, in 1882, he was elected with much ceremony to the
Académie Française. Gradually his health worsened, as a result of
further paralysing strokes. He died in 1895 and, following his state
funeral, was buried in a special chapel built into the basement of the
Institut Pasteur with the walls adorned with phrases symbolising his
achievements in so many fields. That part of the Institut Pasteur is
now a museum to his life and work. The Institut Pasteur has grown
with numerous new buildings to straddle the rue du Dr Roux, named
of course after Émile Roux, with satellite institutes in other cities, and
is a name of world renown.

Pasteur's iconic status was established during his life and his name
is venerated throughout France. There are numerous statues of him,
in addition to his name being used for streets and buildings. An opin-
ion poll of schoolchildren in the 1960s named Pasteur as the person
who had done most for France, with about 50 per cent of the vote. He
is revered scientifically as well. The centenary of the Institut Pasteur
was celebrated nationally and internationally, with a landmark scientific conference where the French President and Prime Minister
performed the opening and closing ceremonies. Pasteur has been
commemorated on stamps worldwide and was the subject of an
Oscar-winning Hollywood film in 1936, *The Story of Louis Pasteur*.[13]

The stunning success of the anthrax work opened the way for a proper scientific onslaught on the cause of other diseases. Koch had shown how to identify bacteria responsible for one disease. There followed a major effort to find bacteria involved in other diseases. Koch himself next studied wound infections using experimental animals to identify several different types of bacteria, including the streptococcus bacteria (streptococci) later found in boils by Pasteur. The human connection was made by the Scottish surgeon Sir Alexander Ogston, who found not only streptococci but another species that he named *Staphylococcus*.

Koch's animal experiments logically linked a particular bacterium with a particular disease. Bacteria isolated from a diseased animal were grown in the laboratory a number of times until he could be sure that there was no carryover from the original sample. These pure cultures of bacteria were then injected into healthy animals to see whether they would produce the same disease. The principles behind this careful approach to investigating disease came to be known as 'Koch's postulates'.[14] They are valuable ideas, but cannot be applied strictly to many diseases (such as those caused by viruses), or indeed to diphtheria and cholera which were investigated in Koch's time. The first description of the postulates was given in a paper by Friedrich Loeffler on diphtheria, whereas Koch himself first noted them seven years later in his work on tuberculosis (TB). Neither, however, wrote any of the precise versions that appear in microbiology textbooks, which probably explains why each textbook states them slightly differently.

In 1880 Koch had moved to a newly built laboratory in Berlin where he could work with the help of assistants in the laboratories of the Imperial Health Office. Then, in 1885 he became Professor of Hygiene at the University of Berlin. Over the years, an impressive list of collaborators worked with him and took his ideas forward, in a similar manner to the French school under Pasteur's leadership. Initially he was joined by Georg Gaffky and Loeffler and, later, Shibasaburo Kitasato, Paul Ehrlich, and others, who would join the microbiology 'hall of fame', arrived in Berlin to work with him.

Koch's greatest triumph was the discovery of *Mycobacterium tuberculosis*, the bacterium responsible for TB, then a major killer, as it still is in the developing world today. This difficult and brilliant piece of work[15] was the peak in his career, but unfortunately it also led later to his considerable discredit. Following his discovery of *M. tuberculosis*, Koch made a preparation from the TB bacilli called tuberculin and wrongly claimed it as a cure, partly as a result of excessive pressure to report his results before he was certain of them. Although tuberculin later proved to be useful as a test for tuberculosis, it was not a cure. In addition, his incorrect but dogmatically stated view that TB in cattle (caused by *M. bovis*) could not infect humans delayed the introduction of pasteurisation that could have prevented it.

During the unfortunate episode with TB, Koch turned his attention to cholera. A further epidemic had reached Egypt in 1883, and France and Germany were each asked to see whether their new science of microbiology could help to understand and treat the disease. Each country sent an expedition—each government wanted to try to tackle the epidemic before it reached its own people. The German expedition was led by Koch himself and included Gaffky. The 61-year-old Pasteur sent Émile Roux and Louis Thuillier as team leaders. Neither team was successful in isolating cholera bacteria. Tragedy struck when the 27-year-old Thuillier inexplicably contracted cholera, although he had apparently not been near a cholera victim for 14 days. He died.

The French expedition returned home, while Koch, ever intent on travelling, and his group journeyed on to India where they stayed for four months. There they isolated the 'comma bacillus', which had previously been found in cholera victims by Pacini and others. In addition the German team conducted some epidemiological studies showing that cholera was linked to infected water supplies—yet further support for John Snow's ideas on how cholera was spread. They returned home in triumph and Koch was presented with a medal by Kaiser Otto von Bismarck. There were still some who did not believe either in the infectious cause of cholera or that Koch had found the bacterium responsible.[16] Part of the problem was that Koch was unable to fulfil his own postulates, because the bacteria did not affect animals, so he could not show that his pure cultures caused disease. This difficulty with his own postulates was thrown back at him in an anonymous article in the *British Medical Journal*. However, others showed that it was possible to cause a cholera-like disease by injecting the comma bacterium straight into the guts of some animals, and this work helped to support his conclusions.

The British government also sent commissions to both Egypt and India, although their purpose was political. Sir Joseph Frayer, who was President of the Medical Board of the India Office, stated in a memorandum to a junior minister in the India Office[17]:

I am also very anxious to avert the evil consequences that may accrue from the effects of this so-called discovery [Koch's germ theory of cholera] on our sea traffic and international communication.

Edward Klein, later on described as the father of British bacteriology, led the Commission to India, and the subsequent report centred on the infectious aspects of cholera. As the miasma theory could no longer be countenanced because of the weight of evidence against it, two other theories remained: contagion and some aspects of local conditions (a view held by von Pettenkofer). Taking the narrow view of infection as occurring from person to person they observed that those nursing cholera victims seldom came down with the disease, which is of course because the common route of infection is by contaminated water, as shown years earlier by Snow. Whereas Koch had shown that some villagers drinking water from contaminated water tanks contracted the disease, the British report emphasised that some had not.

The background to this peculiar situation included the expanding British imperial interests (British rule in India and its virtual control of Egypt), the opening of the Suez Canal, and a disagreement on how to control cholera. The Suez Canal had given the British, in particular, a great trade advantage by shortening the time for travel between Britain and India, and they were keen to protect these interests and prevent any measure that might block their ability to trade. Quarantine, which was promoted by other European countries as the means to control cholera, was seen to be likely to block the flow of trade and injure Britain more than other countries. The British with their improved sanitation measures saw little virtue in quarantine. In the case of cholera they were probably correct, and Koch himself supported the view that impure water and inadequate sanitation were of most importance. It was widely held in the rest of Europe that the outbreak of cholera in Egypt, and subsequently in southern France, had come from India on British trading ships via the Suez Canal. It was seen, particularly in Germany, that political capital could be made from this.

Timothy Lewis, who had studied with von Pettenkofer, was sent to Marseilles to investigate the outbreak there and poured scorn on Koch's findings. Lewis was then appointed by the Secretary of State for India, Randolph Churchill—Winston's father—as secretary of a committee to consider the report of the British Cholera Commission to India. The eventual report, incredibly entitled 'The Official Refutation of Dr Robert Koch's Theory of Cholera and Commas' was published in an obscure journal, which was strange considering the importance of the topic in 1886. Its conclusions that Koch's identification of the comma bacterium as the agent of cholera was 'questionable' and that quarantine was useless were a considerable over-interpretation of Klein's report and were not likely to gain scientific acceptance.

Eight years later an outbreak of cholera in Germany gave Koch the opportunity to finalise the debate. Two neighbouring cities, Altona and Hamburg, each drew water from the same river. Altona's water supply was collected downstream of the Hamburg water supply and therefore was potentially more heavily contaminated with sewage. Surprisingly there was hardly any cholera in this city compared with bad outbreaks in Hamburg. Koch showed that Altona's water filtration system protected its citizens. In many ways this was similar to the earlier work of Snow and John Simon in London, but there were two important differences. First, Koch was able to bring microbiological analysis to the problem—he could show the presence of the causative bacterium to back up the epidemiological evidence. Second, his authority in the field added weight to his conclusions. Cholera in a doctor who accidentally swallowed the comma bacteria on one of Koch's bacteriology courses, at a time when there was no cholera in Germany, was further proof that this was the infectious agent. It was around this time that cholera claimed the life of the Russian composer Peter Tchaikovsky, who drank unboiled water during a cholera epidemic and died at the untimely age of 53.

The cholera work was Koch's last major contribution. At the age of 46, he had become infatuated with a sexy 17-year-old actress, Hedwig Freiberg, with whom he first lived and whom he married three years later in 1891 after an acrimonious divorce. Such an event nowadays would be the subject of amused discussion at scientific meetings. In the early 1890s it was the talk and scandal of German society and at

3.3 Hedwig Freiberg in 1889, aged 17, when she first met Robert Koch.
(© Robert-Koch-Museum im Institut für Mikrobiologie und Hygiene der Charité, Berlin, Germany.)

international conferences. Koch's frequent travelling after this time may have been to escape some of the controversy. He visited India again in 1897 to study plague, and travelled to the recently set up Imperial Bacteriology Laboratory[18] in the foothills of the Himalayas. He also visited Africa, America, and Japan. Koch was reportedly fascinated by Japanese culture. Kitasato, his host during the visit in 1908, is alleged to have arranged for a girl called Ohanasan to go to Germany with Koch.

Much of the rest of Koch's professional life was devoted to non-bacterial diseases, none of which produced particularly ground-breaking results, although his discovery of typhoid carriers, who carried and excreted the bacteria without showing signs of disease, was an important finding. Over the years Koch accrued the highest honours and worldwide recognition,* in spite of the regrettable controversies that some of his ideas had stirred: foreign membership of the Royal Society in 1897 and eventually, in 1905, the Nobel Prize for Medicine. He died in 1910.

Part of the legacy left by Pasteur and Koch was a new enthusiasm and confidence in tackling infectious diseases, as well as a set of powerful and successful principles with which to investigate them. Their disciples carried their work forward, identifying many more infectious causes of disease, including viruses at the start of the twentieth century. The years from 1876 to the start of the First World War saw the isolation of many of the bacteria involved in major disease (Table 3.1). These include the salmonella group that cause typhoid and food poisoning, and the highly related and very versatile group of bacteria (*Escherichia* species), named after Escherich, that cause several different diseases and make an equally versatile range of toxins. The clostridium bacteria that make neurotoxins, *Clostridium tetani* and *C. botulinum*, were also discovered at this time, as was *C. perfringens* (originally named *C. welchii*), a related bacterium that caused gas gangrene to deadly effect in the First World War.

The discovery of the agent of the plague that had caused so many deaths was also part of this success story. There are competing claims about who isolated it. This is often diplomatically avoided by microbiology textbooks, which name both Alexandre Yersin and Shibasaburo Kitasato. Although the bacterium is named after Yersin, some Japanese accounts completely ignore Yersin and heap praise on Kitasato.† Kitasato had worked with Koch for over three years and had been the first to grow *C. tetani* in pure culture. He extended this work to produce immunity to tetanus, in parallel work on diphtheria by von Behring, which was published jointly. Yersin, who had carried

---

*He even had a sailing ship, the *Professor Koch*, named after him. This was launched at Port Glasgow in 1891.

†Nevertheless, Kitasato was not well treated on his return from Germany to Japan in 1883, and he found it difficult to find somewhere to work.

Table 3.1. Bacterial diseases identified in the nineteenth century[a]

Year	Disease	Bacterium	Discoverer	Toxins?
1876	Anthrax	*Bacillus anthracis*	Robert Koch	Yes
1879	Gonorrhoea	*Neisseria gonorrhoea*	Albert Neisser	?
1880	Typhoid	*Salmonella typhi*	Carl Joseph Eberth, Georg Gaffky[b]	Yes
1880	Fowl cholera	*Pasteurella multocida*	Louis Pasteur[c]	Yes
1881	Wound infections	*Staphylococcus aureus*	Alexander Ogston[d]	Yes
1882	Tuberculosis	*Mycobacterium tuberculosis*	Robert Koch	Yes
1882	Skin wounds	*Pseudomonas aeruginosa*	Carl Gessard	Yes
1883	Necrotising fasciitis, scarlet fever, etc.	*Streptococcus pyogenes*	Frederick Fehleisen	Yes
1884	Diphtheria	*Corynebacterium diphtheriae*	Friedrich Loeffler	Yes
1884	Cholera	*Vibrio cholerae*	Robert Koch	Yes
1884	Tetanus	*Clostridium tetani*	Arthur Nicholaier[e]	Yes
1885	Diarrhoea[f]	*Escherichia coli*	Theodor Escherich	Yes
1886	Pneumonia	*Streptoccocus pneumoniae*	Albert Fraenkel	Yes
1886	Swine plague	*Salmonella choleraesuis*	Daniel Salmon and Theobald Smith	Yes
1887	Malta fever	*Brucella melitensis*	David Bruce	Yes
1887	Meningitis	*Neisseria meningitides*	Anton Weischselbaum	Yes
1888	Food poisoning	*Salmonella enteritidis*	August Gaertner	Yes
1892	Gas gangrene	*Clostridium perfringens (welchii)*	William Welch and George Nuttall	Yes
1892	Fever, septic shock	Endotoxin[g]	Richard Pfeiffer	N/A

Year	Disease	Species	Discoverer	
1894	Plague	*Yersinia pestis*	Alexandre Yersin[h]	Yes
1897	Botulism	*Clostridium botulinum*	Emile-Pierre van Ermengem	Yes
1898	Dysentery[i]	*Shigella dysenteriae*	Kiyoshi Shiga	Yes
1905	Syphilis	*Treponema pallidum*	Fritz Schaudinn and Erich Hoffmann	No
1906	Whooping cough	*Bordetella pertussis*	Jules Bordet and Octave Gengou	Yes
1907	Rocky mountain spotted fever	*Rickettsia rickettsii*	Howard Ricketts	No
1909	Typhus	*Rickettsia prowazekii*	Howard Ricketts[k]	?
1911	Tularaemia	*Francisella tularaemia*	George McCoy and Charles Chapin	?

[a] Taking a loose definition of the nineteenth century up to the First World War.

[b] Eberth saw *S. typhi* in the microscope, but it was first cultivated by Gaffky in 1884.

[c] These bacteria were identified in 1879 by Edoardo Perroncito, a veterinarian at the University of Turin, and then worked on by Jean-Joseph Henri Toussaint.

[d] Various bacteria infect wounds. Koch observed bacteria in pus in 1878. These included streptococcus bacteria (streptococci), but Ogston also recognised a new type that he named *Staphylococcus*.

[e] Nicholaier had identified the bacteria, but Kitasato first grew them in pure culture because he realised that they were anaerobic and would grow only when air was excluded.

[f] Other bacteria, and many viruses, can also cause diarrhoea, and conversely *E. coli* can cause other diseases.

[g] Endotoxin is produced by a whole range of bacteria. It is not a protein and therefore has very different properties to the protein toxins.

[h] Some sources credit Kitasato with this discovery, or both Yersin and Kitasato (see text).

[i] Dysentery can also be caused by other pathogens.

[j] *Rickettsia* species are very small bacteria that are called obligate intracellular parasites, that is they have to live inside our cells.

[k] Ricketts is thought to have died from typhus shortly after identifying its cause. Charles Jules Henri Nicolle, a Pastorian, identified the role of lice in transmission, and won the Nobel Prize for this work.

out vital work on diphtheria with Pasteur's group, had decided to escape from Paris and to emigrate to Vietnam as a ship's doctor and explorer, when the violent outbreak of the plague in Hong Kong occurred. The 30-year-old Yersin was sent by the French Colonial Medical Corps to investigate. However, on his arrival he discovered that Kitasato was already there and had been given the full support of the British authorities. They prevented Yersin from having access to the many corpses of plague victims. Moreover, Kitasato had sole use of the only laboratory facilities. Yersin established his laboratory in a makeshift shack with the help of a local priest, who also suggested that he bribe some English sailors who were charged with burying the dead. They took him to the cellar where the bodies were being stored and he was able to cut open one of the characteristic buboes,* which Yersin reasoned would contain the infectious agent. From these he was able to isolate a bacterium that did not take up the Gram stain (see below) and which caused plague when injected into mice. Furthermore, he was struck by the large number of dead rats in the streets and was able to isolate the same bacterium from them. Kitasato, meanwhile, had isolated an organism, not from the buboes, that took up the Gram stain, and had mistakenly rushed into print with the wrong result. The final part of this story was completed by Paul-Louis Simond in 1898, who discovered the role of the rat flea in the life cycle of plague.

One reason for Koch's success was his work with stains that aided the visualisation of bacteria in the microscope and also helped to distinguish between different bacteria. This work was continued, in particular, by Paul Ehrlich who introduced the dye methylene blue.[19] Later Christian Gram, a Danish microbiologist who worked in Loeffler's laboratory in Berlin, discovered the stain crystal violet, which would have a profound effect on microbiology. The Gram stain, first described in 1884, was taken up by some bacteria, but not by others. This deceptively simple classification of bacteria into Gram positive and Gram negative continues to be extraordinarily useful.[20] A further technological advance was the design by Julius Petri in 1887 of a dish, with an overhanging lid for growing bacteria. This dish that now bears his name allowed air to enter but kept airborne bacteria out—a simple but important advance still in use today.

---

*Large black swellings, one of the main symptoms of the disease.

The new science of microbiology spread beyond France and Germany. Kitasato took it back to Japan, when he returned there after several years of working in Koch's laboratory. Other Japanese workers included Kiyoshi Shiga who had worked with Ehrlich, as had Sahachiro Hata. Britain had already been exposed to microbiology—after all it was in London that Pasteur and Koch had first met in 1881 at the International Congress of Medicine. Moreover, Lister was a significant player in the development of the science of microbiology and Tyndall had played an important part in the defeat of spontaneous generation theories. In 1887, Sir David Bruce, a British army doctor, identified the bacterium responsible for Malta fever, now named *Brucella melitensis* after him (and Malta, where he identified it). He later worked on other diseases including sleeping sickness. Sir Almwroth Wright, the so-called 'father of English bacteriology', published the first attempt at a typhoid vaccine in 1897.

The debate about spontaneous generation and the germ theory of life had attracted interest in the USA. However, it was in the 1880s that interest became widespread, when the clarity of the work by Pasteur and Koch could not be ignored. Many Americans who had been trained by attending one of Koch's courses returned home to use these methods in newly created microbiology laboratories. These young enthusiasts started to make their mark to break the European stranglehold on the new discipline. For example, Daniel Salmon, a Cornell-educated veterinarian, identified the cause of swine plague and was immortalised in the naming of the genus, *Salmonella*. In 1899, the formation of the American Society for Microbiology marked the maturation of the new science in America.

The long and gradual accretion of knowledge about bacteria and their possible role had come to an exciting climax, in the late nineteenth century, to define the new science of microbiology. Indeed the science of immunology[21] was also begun through this work. There was a great hope then that humans could quickly conquer infection in the same way that they appeared to be able to control other aspects of Nature. However, the great advances in combating some diseases were counterbalanced by failure in tackling others. The genius and boldness of Pasteur, Koch, and the other less well-known figures of the late nineteenth century had set microbiology on a firm footing, but there was no real understanding of the science of these diseases or

of how the infectious organisms interacted with their unwilling hosts. The next stage would unravel how bacteria, and in particular toxins, worked against us and would explain not only how infection worked, but surprisingly also show how we worked.

3.4 (*opposite*) Robert Koch in India, at the Imperial Bacteriology Laboratory in Mukteswar in 1897. Sitting (from left to right): Alfred Lingard (the Imperial Bacteriologist), Robert Koch, Richard Pfeiffer, and Georg Gaffky. Second from the left at the back is Major FSH Baldrey, who worked on haemorrhagic septicaemia. The others are not known.

(The photograph was given to the author when he visited the Indian Veterinary Institute in 1988.)

4

THE ANATOMY OF DIPHTHERIA
Taming the deadly scourge of childhood

Diphtheria occupies a particularly special place in microbiology and its history. It has probably been studied in more depth than any other disease. We know exactly how it causes damage to our bodies, which surprisingly is still not known for most diseases. Diphtheria toxin was the first complex toxin to be understood at the molecular level, and it continues to produce interesting results. It represents the ultimate model toxin and illustrates beautifully how such toxins work. Its malevolent action was one of the first to be tamed by vaccines so effective that the basic formula has not changed substantially in over 80 years. In addition, its potent action has now even been turned to cancer therapy.

Diphtheria is no longer a big health problem in the west—in the USA there were 49 cases recorded in the 20 years from 1980 to 1999. The situation in the supposed developed world was not always like this—100 years ago, diphtheria was referred to as 'the deadly scourge of childhood'. Many stories of individual family tragedy have been recorded that show how terrifying both the disease itself and the prospect of catching it were. Even in the 1920s this was a dreadful disease.* In 1921 there were 206,000 cases with 15,500 deaths in the USA alone. Diphtheria recurred in Europe during the Second World War with about a million cases. Our present relative complacency about this distressing disease contrasts with the situation in the 1990s in the countries of the former Soviet Union, where in the Russian Federation there were more than 110,000 cases and 2,900 deaths.

*The word 'diphtheria' still carries a dreadful reverence. One heavy metal band has chosen it for their name. There is also a band called 'Anthrax', but I am not aware of any others with toxin-related names.

Diphtheria is a striking disease because of its symptoms and effects. It is one of the fastest acting bacterial diseases, often killing within a week of the start of the symptoms. It begins with a sore throat and slight fever, but then the ensuing destruction of the lining of the throat rapidly induces formation of the main characteristic of the disease, the diphtheric* membrane. This tough leathery film forms across the throat and is made from dead cells, liquid oozing from the damaged throat, and bacteria. By itself it can lead to one dreadful form of death—suffocation, because the victim is unable to breathe. However, even if patients can breathe, they are likely to die from the severe injury that the toxin causes to other organs in the body. These include the heart, other muscles, the liver, and the kidneys.[1] In other words the whole body is ravaged by the disease, but bacteria are found only in the throat. It was this realisation that led to the idea of a poison, or toxin, that coursed around the body.

Diphtheria has been given many names over the centuries. These include croup, angina maligna (because of the damage to the heart), malignant ulcerous sore throat, and throat distemper. In Spain, because of its murderous suffocating death, it was called morbus sufficans or garotillo (the strangler).

It is usual to grant Hippocrates the distinction of the earliest description of diphtheria. Indeed Hippocrates is still held in such esteem that he tends to be attributed with every early medical description. His report credited as a portrayal of diphtheria is very vague. A much better picture was provided by Areteus, a Greek doctor of the second century AD[2]:

Ulcers occur on the tonsils ... some ... mild and innocuous; but others of an unusual kind, pestilential and fatal ... the inflammation seizes the neck; and these die within a few days from the inflammation, fever, foetid smell, and want of food ... if it spread to the thorax by the windpipe, it occasions death by suffocation within the space of a day.

He expresses the view that the disease is caused by swallowing cold or hot substances, and different 'airs'. His description of death by this unpleasant disease is particularly chilling:

The manner of death is most piteous: pain is sharp and hot ... respiration bad, for their breath smells strongly of putrefaction ... they are in so loathsome

---

*From the Greek word for leather.

a state that they cannot endure the smell of themselves ... fever acute, thirst as from fire, and yet they do not desire drink for fear of the pains it would occasion; for they become sick if it compress the tonsils, or if it return by the nostrils ... these symptoms hurry on from bad to worse, until suddenly falling to the ground they expire.

Other descriptions followed over the centuries and gradually teased out important aspects of the disease. Guillaume de Baillou in Paris was among the first to describe the characteristic membrane in 1576. He also realised the potential value to the patient of tracheotomy, where an opening is made in the throat to allow breathing. In those days, it was sometimes carried out with a sword! Such crude 'operations' were not without their risks, but any option was worth trying if the patient was going to suffocate.

The description of diphtheria in a pamphlet by the Englishman John Fothergill did not bring any special new insights to our understanding of diphtheria, but it raised interest in the topic. Born in 1712 in Carr End in Wensleydale, Fothergill studied at Edinburgh University, then at St Thomas' Hospital in London. His 'account of the sore throat attended with ulcers' was an early English description of the disease. We now know that it mixes up a description of two diseases—scarlet fever* and diphtheria—but at the time it earned him both fame and fortune. Despite this, he reportedly was more than willing to treat the poor, and he was excessively hard working. His American friend Benjamin Franklin† once wrote to him 'By the way, when do you intend to live?' He also said of Fothergill 'I can hardly conceive that a better man ever existed'. Fothergill helped Franklin to establish the Pennsylvania Hospital in America. Fothergill also wrote the first descriptions of coronary arteriosclerosis (hardening of the arteries), angina, and migraine. These are probably greater scientific achievements. Besides all this, he was a well-known botanist. He initiated the collection of plants (over 3,000) for their potential medicinal value and ran a botanical garden at Upton House in London. Various artists were commissioned to draw the plants. Some of this unique collection of pictures was bought after his death by Catherine the

*Scarlet fever is one of the diseases caused by *Streptococcus pyogenes*, a bacterium that makes a battery of toxins.

†Franklin, famous for his work on electricity, also helped to draft the American Constitution.

Great of Russia, and has just recently been re-discovered. Fothergill was a committed Quaker, active against slavery, and a prison reformer. He was a founder of Ackworth School in Yorkshire, where the school hall is named after him. As well as all his other achievements, Dr Fothergill is also credited with the increased popularity of coffee drinking in England.

There were other doctors who described the disease around this time. The Englishman John Huxham, as well as describing diphtheria, also wrote about scurvy. He was a colourful character, with a scarlet coat and gold-headed cane. Apparently he tried to drum up support from prospective patients by falsely appearing to be in great demand. He would get his servants to summon him urgently from church even when no one had requested his assistance, and then he would jump on his horse and gallop off towards the next town. It is not known if this was a successful strategy. The Scotsman Francis Home was not just Professor of Medicine at Edinburgh, but later he was Professor of Agriculture. Home is attributed with the first clear description of the disease, and he noted that it was infectious[2]:

The frequent transmission of the disease to those near by, or to a whole family, by the breath or by the bloody matter ejected from the mouth, proves this to be a contagious disease ...

The American Samuel Bard was also educated at Edinburgh Medical School, but not before surviving capture by the French on his way there. His release was brokered by Fothergill's friend Franklin. While in London en route to Edinburgh, Bard met Fothergill and other famous medical men of the time. After his Scottish education, he returned to New York where he eventually became President of the College of Physicians and Surgeons. He also recognised the infectious nature of diphtheria.

The next real advance in understanding came in 1826, about 50 years later, when Pierre Bretonneau* defined the disease based on the diphtheric membrane in the throat. Bretonneau was the first to use this distinctive feature as a definition of the disease. Until that time the slightly different symptoms of the disease in different people had led to argument over whether doctors were dealing with the same

*Bretonneau came from a long line of surgeons, but because of ill-health only became fully medically qualified at the age of 36 in 1814.

disease. This problem is typical of many diseases—the same common cold can affect people even in the same family in different ways, and conversely deaths due to AIDS take many different forms, although all can be linked to HIV.

Bretonneau was strongly in favour of tracheotomy to enable the patient to breathe and he recognised that diphtheria was infectious. He also investigated typhoid fever and realised that each infectious disease was specific and different. This last point led on naturally to the idea that each disease had its own specific cause and anticipated the great period of microbiology at the end of the nineteenth century.

The story reached its initial climax during the last three decades of the nineteenth century. Diphtheria was conclusively shown to be an infectious disease in 1871. Rabbits inoculated with material from infected humans produced the diphtheric membrane. The hunt was on to find the infectious agent. The 1880s was a very bad time for diphtheria. Children's hospitals were full and the death rate was particularly high. Edwin Klebs claimed, in 1883, to have found that the cause was not one, but two, bacteria, one rod shaped (a bacillus) and one ball shaped (a coccus). The correct answer was provided the following year by Friedrich Loeffler, a disciple and assistant of Robert Koch. He showed that the rods alone caused all the signs of disease in rabbits and pigeons. However, he could isolate the bacillus only in the membrane in the throat. He wondered whether the damage caused throughout the body was the result of a soluble poison. The concept of a bacterial toxin had been born.

There was still some doubt in Loeffler's cautious mind as to whether this bacterium, named *Corynebacterium diphtheriae*, was the cause of the disease—perhaps because of his overzealous interpretation of Koch's postulates. Diphtheria did not fit well with the postulates. The bacteria could be located only in the throats of children who had died of the disease and of experimental animals. This did not follow the pattern of disease throughout the whole body. He had failed to isolate the bacterium from every case of the disease and he had also found it in one child without disease. The issue was settled beyond doubt in 1888, by Émile Roux and Alexandre Yersin, who both worked in Paris at Pasteur's new research institute. They found the same bacterial rods in children with diphtheria. When these bacteria were put into rabbits, any survivors were often paralysed, in a similar fashion to the children who survived. Again they could not find the bacteria anywhere else in the body but the throat. Similar to Loeffler, they argued that the bacteria growing in the throat produced a poison that was released to travel around the body.

Roux and Yersin reasoned that bacteria grown in the laboratory should also make the toxin, which would be released and be present in the liquid broth in which the bacteria had been growing. These experiments, rather breathlessly described in Paul de Kruif's book *The Microbe Hunters*, were potentially very dangerous. The bacteria were removed from the broth by passing the turbid broth full of bacteria through a fine filter and collecting the clear liquid that was free of bacteria. The filters were easily clogged and the broth without bacteria had to be pushed through with air at high pressure. It is alarming to think of the consequences of diphtheria toxin being sprayed around a laboratory at high pressure. However, the first experiments were perfectly safe—the liquid that passed through the filter was not toxic. The characteristic signs of disease were seen only in a final last-ditch experiment when a massive volume of the bacteria-free soup was injected—as de Kruif describes almost too enthusiastically, the poor guinea pig could as easily have drowned as so much liquid was injected under its skin. Why was this supposedly deadly bacterium so weakly poisonous? It turned out that the bacteria had not been given long enough to make toxin.[3] When Roux and Yersin let the growing bacteria stay in the broth for longer, the material passing through the filter was so toxic that a gram would have been enough to kill 25,000 guinea pigs.

As well as its importance for the diphtheria story, this experiment was a key advance in biology. The idea that an inert, non-living, chemical substance could reproduce the signs of a disease was a leap forward in concept, and was further proof that biology is only complicated chemistry.

The next strand of the story switches again to Germany, to Emil von Behring, who worked with Koch in Berlin at the same time as Paul Ehrlich and Shibasaburo Kitasato. Von Behring was searching without much success for a chemical cure for diphtheria. The chemical treatments were often as harmful as diphtheria infections, but some animals survived both the chemical and the bacteriological onslaught. Von Behring wanted to know whether these animals would now be resistant to infection. They were. To see what was happening he mixed up some of the bacteria with the serum* from these animals. He was initially disappointed to find that the bacteria were unaffected by this treatment. Then, von Behring remembered that the disease could be caused by the toxin itself without any bacteria being present. He found that the serum had the ability to inactivate the potent filtered toxin preparations. He showed that the anti-toxin serum given to other animals could make them resistant (immune) to the toxin. Thus the concept of an 'anti-toxin' was established. The demonstration that a non-living substance, that is something in serum, was the cause of immunity enabled the process to be analysed in a more scientific way, for example by purifying the substance responsible. This work was published in 1890 jointly with Kitasato's work on tetanus, and showed that treatment of both diseases was possible:

These properties are of such a lasting nature that they remain effective even in the bodies of other animals, and thus we are in a position, by means of blood or serum transfusion, to achieve excellent therapeutical effects.

Rather poetically—for it is thought that von Behring was a bit of a romantic—this incredibly short, but highly significant paper ends with a quote from the poet and scientist Goethe:

Blut ist ein ganz besonder Saft
[Blood is a very special fluid]

It has been suggested that the first child to be treated with anti-toxin was saved on Christmas Eve 1891, although this may be a fanciful and wistful version of events. Nevertheless anti-toxin treatment did begin about this time. Although the chances of survival were greatly improved, anti-toxin treatment did not represent 100 per cent cure. There were still children dying, in some cases immediately after the anti-toxin was injected into them. Back in Paris, Roux turned his attention to improving the anti-toxin preparations. He made anti-toxin in horses by injecting toxin preparations that had been weakened by adding iodine to them. The horses did not die and produced large quantities of anti-toxin. This new preparation of anti-toxin was

---

*The liquid remaining when the cells are removed from blood. Serum contains antibodies that can recognise foreign proteins such as toxins.

tested at two Paris hospitals. In the Hôpital des Enfant-Malades the death rate among treated children was 24 per cent, whereas at the Hôpital Trousseau, where anti-toxin was not used, it was 60 per cent. The comparison between hospitals was important because mortality from diphtheria varied between outbreaks.

Von Behring saw that there were two reasons why the treatment was not perfect. First, it had to be administered quickly before too much damage had been done to the body. It is now known that the toxin works inside cells where it is resistant to antibodies that cannot cross the cell membrane. Second, the body recognises the horse proteins as foreign and can switch on its immune system to combat these, in some cases catastrophically, leading to shock and death. Von Behring tried to purify the anti-toxin activity from the serum, and showed that the active molecule was a protein. The purified anti-toxin produced fewer side effects because it contained less protein. Thus, von Behring began the use of serum therapy, which in improved form is still used today. His attempts to manufacture anti-toxin against diphtheria on a large scale ran into difficulties that were largely solved by collaboration with Paul Ehrlich, who had worked out how to measure the relative potency of anti-toxins from his work on the plant toxin ricin. They later had a major and bitter disagreement about diphtheria anti-toxin.

The significance of this work with diphtheria cannot be overestimated. A survey of newspaper articles around this time shows an abrupt change in tone. Previously diphtheria had been mentioned in gloomy articles about inquests following diphtheria deaths and descriptions of diphtheria outbreaks. In 1894, the proceedings of the International Hygienic Conference in Budapest were reported daily in *The Times*. Roux's results were described with enthusiasm* and this was followed by numerous other articles and letters, including a letter from Lister describing how Britain would take up the new treatment.†
The brilliant and complementary experiments by the 'two Emils' also helped to dissolve the bitter rivalry between the Pastorian and Kochian camps. Roux and von Behring were jointly honoured in Paris with a prize for their work. Later Roux was godfather to one of

*The Times, 8 September 1894, page 5.
†The Times, 31 October 1894, page 10.

von Behring's sons. Von Behring was fêted with other honours, including a life peerage in Prussia, and in 1901 he received the Nobel Prize for Physiology or Medicine.

Nevertheless this approach was not preventive and the effect of the serum injections was short-lived. Essentially the administered anti-toxin reacted with free diphtheria toxin, but it did not last for long in the body. What was required was a vaccine that could mimic the diph-theria toxin and trick the body into making an immune response, without killing patients. Diphtheria toxin was too toxic to use as a vaccine, and the first attempts to make a safe version that could still raise immunity naturally enough used the anti-toxin. Mixtures of toxin and anti-toxin were suggested as a possible vaccine by the American Theobald Smith in 1907, and tested by von Behring in 1913. The first problem was the great danger of injecting toxin, espe-cially if the ratio of toxin:anti-toxin was not carefully controlled. This could result in the injection of active diphtheria toxin. Second, there remained the possibility of bad reactions to the antibody, which itself was a foreign protein. What was needed was some treatment of the toxin that would render it harmless but still able to stimulate the body's immune system.

The final episode in this part of the story occurred in Britain. In 1923, Alexander Glenny and Barbara Hopkins showed that diph-theria toxin preparations treated with the chemical formaldehyde were no longer toxic, but were able to induce a strong protective immune reaction. Formaldehyde forms bridges between different parts of the protein, and thus wraps the protein in a web that prevents it being able to function properly. A similar finding was published by Roux's nephew Gaston Ramon at the Institut Pasteur. This inactivated mater-ial was called a toxoid, and could be given to healthy people to make them immune to diphtheria. Glenny then showed that the toxoid could be made more effective by further simple chemical treatment.

In 1913, Béla Schick had developed a safe reliable way to test whether someone was already immune to diphtheria. Tiny amounts of toxin were injected under the skin. If the injection site became red and swollen the patient was obviously not able to prevent diphtheria toxin action and had not seen the disease before. However, if there was no obvious reaction the person was immune and vaccination was not required.

Thus by the mid-1920s everything was in place to diagnose, treat, and even prevent this terrible and dangerous disease. However, immunisation was not widely practised in Britain until the 1940s, perhaps because of a reluctance to believe in its effectiveness. Mass immunisation has now almost wiped out the disease. In the Newly Independent States, the collapse of the social order in the immediate aftermath of the break-up of the USSR interrupted normal immunisation. Immunity lasts for only about 15 years and few adults receive booster vaccines, although in some cases immunity may be 'topped up' by occasional exposure to *C. diphtheriae.* So the concern in the west about this outbreak was partly self-interest. Although most adults are probably susceptible to the disease, in a properly immunised society they do not catch diphtheria because they are not exposed to infected children. The prevalence of international travel raised the potential of a serious outbreak in an unprotected adult population. Health officials were therefore worried that the epidemic could spread, and indeed there was a very small increase in the incidence of the disease in adjoining countries. Luckily a substantial immunisation campaign was able to reduce the scale of the disease in Russia, and avoided an even worse tragedy.

However, it is not just with immunisation breakdown that difficulties can arise with diphtheria. An epidemic in Algeria in 1993 killed over 30 people, mainly teenagers and young adults, even though there was a high uptake in immunisation. Some countries now recommend that booster injections be given.

For those unfortunate to succumb to diphtheria, treatment has not really changed since the 1920s. Horse anti-toxin is given to inactivate toxin that is already circulating round the body, but this cannot work on any toxin that has already wormed its way into our cells. Antibiotics are given to kill the bacteria growing in the throat, and the patient needs substantial rest to enable the body to repair itself. Damage to the heart is a major concern. The eventual outcome depends on early diagnosis, and five per cent of those affected die.* Any possible contacts also have to be treated. As well as the human cost there is a financial one too. The World Health Organization has estimated that each case costs about £6,000 to treat.

---

*In the outbreak in the former Soviet Union the mortality rate among young children was 50 per cent in some places.

4.1 Alwin Pappenheimer.
(Photograph supplied by John Collier, Harvard University.)

By 1925 a vaccine had been developed that was so effective and so safe that it is still used today, and has been a miraculous success. The 'deadly scourge of childhood' had within a period of about 50 years been defeated. That could have been the end of the diphtheria story, but in some ways it was only the beginning. No one knew how the vaccine worked or what the diphtheria toxin did that was so deadly. And it is always dangerous to rely on something that you do not understand.

Much of the twentieth century story of diphtheria research revolves round Alwin Max Pappenheimer Junior (1908–1995).[4] Pappenheimer came from an academic biology background, and he and his two siblings all ended up as professors at Harvard. Their father was a well-respected pathologist at Columbia University in New York, and Pappenheimer was surrounded by his father's scientific friends from an early age. Later these contacts encouraged and enabled him to work with bright and able people. This is not to say that his success was the result of unfair patronage and privilege, but it is always the case that potential requires the right environment to be fulfilled.

At Harvard he carried out graduate studies in chemistry but was fascinated by the new subject of biological chemistry. As with Pasteur, this background influenced his future direction. Following his doctorate, he worked for a year on bacterial sugar molecules, and then spent two years working in London. His research in both places had been with bacteria that caused disease. This area had caught his imagination. He later said of this early career that 'it worried me a great deal that I did not have an important problem in mind'. However, on his return to the USA, he had decided that he wanted to devote his scientific career 'to isolate a pure potent bacterial protein toxin and to find out what made it so toxic'. This was a formidable challenge, because, although the idea of a biological poison had been around for about 50 years, it had not been developed to any significant extent. Indeed it would have been impossible to understand the action of diphtheria toxin at that time, as many aspects of cell function were unknown.

Pappenheimer began his research at the Antitoxin and Vaccine Laboratory in Jamaica Plains in Massachusetts in 1935. In the early 1940s he was recruited by Colin MacLeod[5] to the New York University School of Medicine, where he stayed until 1958, apart from war service when he was engaged in the American effort to make vaccines against botulinum toxin. Then, Pappenheimer was headhunted back to Harvard where he stayed until his retirement. He was nevertheless one of those lucky people, whose work was his love—a true amateur—and he remained interested in the progress of both toxin research and his protégés until his death.

The starting point for investigating how the toxin worked was to try to purify it. The toxin was known to be released by bacteria into the complex nutrient medium on which the bacteria were grown. Pappenheimer's strategy was to grow the bacteria in as simple a mixture of foods as possible, in order to make purification easier. He was not alone in pursuing this strategy and the papers of the time show how the advances of one group were built on by others. In early attempts either the toxin was lost during the purification process or the toxin preparations were not active. By 1937 he had devised a successful method that gave a good harvest of active toxin. Indeed, the process he had devised was so good that on one occasion he purified 6 grams of the toxin. As diphtheria toxin kills people at a dose of 0.1 μg (one millionth of a gram) for each kilogram of body weight, his 6 grams was equivalent to almost a million lethal doses. Such an experiment would be unthinkable today without strict safety controls.

During his purification studies, he found that the presence of iron in the growth medium inhibited the production of the toxin. This opened a new field of research, and it is now known that bacteria measure the concentration of dissolved iron around them in order to determine their environment. They then decide how to react. A high iron concentration signals the presence of damaged tissues that are leaking out blood, which has a high iron content. Blood contains nutrients that will support the growth of bacteria and therefore there is no need to produce diphtheria toxin. A low iron concentration is interpreted to suggest that there is little food around for the bacterium to grow on. It switches on production of the toxin, leading to cell damage that will provide nutrients for the bacteria.

The scientific paper that explained in broad terms how diphtheria toxin worked was published in 1958 and, although Pappenheimer was not one of the authors, his influence is obvious from the generous acknowledgement of the authors, Norman Strauss and Edelmira Hendee. Reading old scientific papers is always interesting. Some studies have been so superseded by more recent findings that they appear outdated and irrelevant, whereas others retain a freshness where you can still feel the excitement of the discovery. The Strauss and Hendee paper falls into the latter category.

They were able to use a relatively new advance—the ability to grow human cells in the laboratory. Cells are grown as a single layer in a glass bottle lying on its side with a small volume of growth medium bathing them. The bottle is then kept in a special box (a carbon

dioxide [CO_2] incubator) that enables the cells to be kept at 37°C surrounded with a low concentration of CO_2—in other words to mimic
the conditions in the body. Normal human or animal cells need reassuring signals from other cells to grow and therefore are difficult
to grow in the laboratory. Scientists first had success with cancer
cells that do not require these signals. The cells used by Strauss and
Hendee came from a cancer of the cervix. The cell line, called HeLa
cells, was taken from the unfortunate Henrietta Lacks in 1952 and is
still being used today.

It was known that the toxin had many effects on cells, causing them
to fall apart about four to six hours after diphtheria toxin had been
added. Which, if any, of these effects was the primary event was unknown. Noting that toxin treatment appeared to decrease the amount
of protein in the cell, Strauss and Hendee used a very sensitive
method to look at the formation of new proteins using a radioactive
amino acid precursor. Any new protein made during the experiment
is radioactive. It is easy to separate the large proteins from the relatively small amino acids and thus to measure very accurately how
much new protein has been produced.

The results of the first experiment that they published with this
technique are beautifully clear. Without toxin the cells carried on
making new protein at a steady rate. However, protein synthesis was
stopped dead when diphtheria toxin was added, with less than a millionth of a gram of toxin in each millilitre of the protein synthesis
mix. Moreover, and this was the most exciting part, at the highest
concentration tested the effect occurred within an hour of adding
toxin to the cells. As all the other effects happened much later, the
obvious conclusion was that the key effect of diphtheria toxin on the
cell was to inhibit protein synthesis. The authors expressed it rather
more formally, but still some of the excitement comes across: 'the
inhibition of incorporation of this amino acid by toxin is striking and
abruptly follows a period of normal uptake.' They clearly recognised
that the binding of toxin to the cell and the toxic effect were two different processes, but they also pointed out that it was not known
whether the toxin bound to the outside of the cells or got inside.* By

---

*The lag time between the addition of toxin and the inhibition of protein synthesis
reflected the time taken for the toxin to get inside the cell and reach its target.

any standard this was an exciting piece of work. This was the biggest advance in understanding how diphtheria toxin affected the cell since its discovery over 70 years earlier by Roux and Yersin.

Of course lots of questions remained, including the key one—to work out the molecular action of the toxin. It was difficult to try to identify the primary molecular event in the whole cell, which, although simpler than a whole animal, is a highly intricate mass of interacting processes. Protein synthesis itself is a complex process with many different stages, so there were many steps where the block could have occurred. An even simpler technique than the HeLa cell system was needed. The process of protein synthesis was reconstructed in a test tube, which allowed the individual stages to be looked at. John Collier, working with Pappenheimer, found that the obstruction was where new amino acids were joined to the growing protein chain. Adding toxin to this in vitro* system inhibited protein synthesis. When the toxin preparation was treated with antitoxin, the inhibiting activity was also removed, showing that the inhibitory effect was due to diphtheria toxin and not some unknown contaminating substance.

The next clue as to what was happening came in the same paper. As is so often the case, serendipity played a part. As the HeLa cell testtube system was very expensive and time-consuming, they turned to another cell-free system that others had recently developed. This was formed using rabbit red blood cells that had been broken open. They were astonished to find that the toxin did not work in this new system because they knew that rabbits were sensitive to diphtheria toxin. Rather than abandon this as a failure, they set out to find out why. In these in vitro systems, the protein-making machines, the large ribosomes,[6] can be separated from the rest of the smaller, more soluble components, and so they tried out mix-and-match experiments using ribosomes from either the HeLa or the rabbit cell system with the smaller molecules from the other system. The rabbit ribosomes were indeed very sensitive to inhibition by diphtheria toxin, but only if the smaller molecules from the HeLa cells were present. The component that was responsible for this effect was smaller than a protein. It could survive being boiled for 10 minutes, which was further evidence that

*In vitro means 'in glass', although nowadays plastic is more likely to be used.

it was not a protein, as most proteins are very sensitive to heating. They then tried out a number of chemicals that were commercially available and lying around in their laboratory. They found that adding minute amounts of nicotinamide adenine dinucleotide (NAD)* to their rabbit protein synthesis system made it exquisitely sensitive to diphtheria toxin. All sorts of other explanations were carefully ruled out. In case the NAD was contaminated, they showed that an enzyme that specifically destroyed NAD also destroyed the effect. Finally, in this landmark paper they showed that the step that was inhibited was the addition of the amino acid to the growing chain of the protein.

Over the next few years it was shown that a protein involved in the process of transferring the amino acid to the growing chain was inactivated by diphtheria toxin. This protein was called elongation factor 2 (EF-2). Then, in 1968, Tasuku Honjo, working in Kato's laboratory in Tokyo, and, independently in 1969, Michael Gill, working with Pappenheimer, discovered that NAD was cut into two pieces by the action of the diphtheria toxin. Part of the molecule, called ADP-ribose, was attached to EF-2. The addition of ADP-ribose to EF-2 clearly upsets the ability of this important protein to do its job. Diphtheria toxin is an enzyme that facilitates the chemical reaction linking ADP-ribose to a molecule of EF-2. As the toxin is not destroyed in this reaction, it can then attack and inactivate many more copies of EF-2. Indeed it is known that only one molecule of the toxin is needed to kill a cell. Like all good crime stories, everything seems very obvious and straightforward when everything has been explained. It is not surprising that diphtheria toxin is so potent and causes so much damage to the body. It assaults a crucial cellular function and each molecule of the toxin is able to attack several molecules of EF-2.

Although the basic facts of diphtheria toxin action were now known, diphtheria still had some surprises in store. The gene that specifies the information for diphtheria toxin was not found in all *C. diphtheriae* bacteria. Harmless bacteria became deadly only when they were infected with a bacteriophage† that carried the gene for

---

*NAD is a molecule that the cell uses to shuffle around hydrogen ions. It is made from two nucleotides (building blocks for DNA and the genes).

†Bacteria can be attacked by viruses known as bacteriophages. These viruses often carry genes involved in disease.

diphtheria toxin. An extra surprise was that the toxin gene was controlled by bacterial and not the viral genes—an early example of interaction between two different types of biological organism. In 1983, the gene for diphtheria toxin was isolated, which allowed scientists to work out the precise sequence of the amino acids that make diphtheria toxin what it is. Thus a biological poison had been reduced to a defined chemical molecule—certainly a complex molecule, but one that obeyed the conventional rules of chemistry, and not some mystical or unfathomable force.

The lag period after toxin is added to cells, seen in the landmark experiments on protein synthesis, was attributed to the entry process. How toxin entry occurred was worked out over a number of years. The key thing is that the toxin has to do three things. First, it has to attach strongly to cells, next, it has to get across the cell membrane, and, last, it has to execute its potent action on EF-2. It eventually turned out that there were three distinct parts, or domains, of the molecule, each of which carries out one of these functions.

Diphtheria toxin can be split easily into two pieces—John Collier showed that the A (or active) fragment attacks EF-2. The remaining fragment, not surprisingly called the B (or binding) fragment, deals with the other two functions (toxin attachment and crossing the membrane). All toxins that get into our cells follow this basic AB pattern. In diphtheria toxin, the A and B parts are held together by a loop of amino acids that can be cut by enzymes called proteases. However, the toxin stays together until after cell entry, when the active region is released. In other words the chemistry enables the A fragment to be released when it needs to be.

Diphtheria toxin binds on to the surface of the cell, and gets taken inside in a little bag of membrane called a vesicle, which is pinched off from the main cell membrane. By this stage the toxin is inside the cell, but has not yet succeeded in penetrating across the membrane to reach the main compartment of the cell. The B fragment was shown by Patrice Boquet, working with Pappenheimer, to be hydrophobic (water hating or fat liking).[7] This fitted with its ability to penetrate through the membrane of the cell, which is not a trivial operation, because a basic function of membranes is to act as a barrier to keep the content of a cell inside. Naturally they are resistant to proteins drifting across them. The vesicle is trafficked (moved) through the cell and this takes some time to happen, which of course explains the delay before protein synthesis is stopped. Experiments to understand toxin/membrane interaction in detail using diphtheria toxin as a model system are still being published, in particular by the Norwegian Sjur Olsnes, who also worked at one time with Pappenheimer.

The ultimate explanation of the chemistry of diphtheria toxin was provided by the three-dimensional picture of the toxin[8] produced in 1992 by Senyon Choe and colleagues. The crystal structure confirmed the structure of the diphtheria toxin that had been worked out using other techniques. In particular it was possible to discern the three independent domains that carry out the three functions of the toxin, and also the exposed loop that joins the active A fragment to the rest of the molecule, and which is susceptible to enzymatic cleavage.

The transmembrane, or translocation, domain is perhaps of greatest interest, because it explains elegantly how the toxin uses the environment of the cell to its advantage to be able to force the active domain across the membrane. The transmembrane domain is made mainly of bundles of helices, with the amino acids arranged in coils. Some of these helices are hydrophilic whereas others are hydrophobic, and therefore eager to enter fat-based substances such as the inside of membranes. However, there are several acidic amino acids in the loops between the tips of these hydrophobic helices. These are hydrophilic at neutral pH* because they are electrically charged. As the little vesicle with the toxin in it is trafficked through the cell it gradually becomes slightly more acidic—not as acidic as car battery acid, more like vinegar. As there is an increased concentration of hydrogen ions (charged hydrogen atoms) in an acid, this relatively mild pH change causes hydrogen ions to become attached to the acidic amino acids to neutralise the charge. The amino acids in the loops and helices are all hydrophobic now, and can stab into the cell membrane to drag the A fragment across it. Slightly acidic conditions also enable the bond holding the A and B fragments together to break, and thus release the A fragment. These acidic conditions are exactly the conditions that the diphtheria toxin finds itself in about an

---

*pH is how acidity or alkalinity is measured. A pH of 1 is a strong acid, a pH of 7 is neutral, and a pH of 14 is a strong alkali.

hour after it has been brought into the cell in a vesicle, and they enable it to escape to do its work.

So over a period of 100 years not only was diphtheria conquered, but it was one of the first diseases to be understood in molecular terms, indeed even in atomic terms, because we can see precisely how the chemistry explains the biology of this clever molecule.

What is there left to say about diphtheria? The recent outbreaks in the former Soviet Union show diphtheria's continuing potential to cause fatal disease rather chillingly. However, diphtheria has so far not managed to come up with an answer to the vaccine—we have to hope it stays like that. Why has the vaccine been so effective, and why has the bacterium so far failed to fight back? First, the course of the disease, its extreme severity, depends crucially on the toxin. So, when that weapon is incapacitated, the virulent bacterium is incapacitated. In addition, the toxin can be hunted down by antibodies because it is released from the bacterium and then travels alone as it ravages the body. The diphtheria toxoid induces us to make several different antibodies that bind to the toxin in different places, each of which can prevent the toxin from binding. Presumably, the multiple ways of inhibiting the toxin action have made it difficult for diphtheria to fight back by making changes that would avoid this antibody attack and yet still enable the toxin to bind to our cells.

Diphtheria has shown us a model not only for the way toxins work, but also for how proteins work. The present importance of that insight cannot be over-estimated, at a time when scientific interest is switching from the genes to the proteins—the workers of the cell. The potent diphtheria activity has also been harnessed to kill cancer cells, as is discussed later. Furthermore, other uses have been found for diphtheria toxin in basic biology, where it can be used to help us to understand how the single fertilised egg changes to a fully formed multicellular human. So the deadly scourge of childhood appears to have been tamed and has also paid back a little of its debt to humankind.

5

UNDERSTANDING

All toxins fit into three basic mechanisms

There are hundreds of different bacterial protein toxins. Early attempts to classify them relied on looking at the types of diseases that they caused. The neurotoxins (that caused problems linked to the nervous system) included botulinum and tetanus, but also diphtheria, which can induce nerve damage. Enterotoxins were those that caused enteritis, for example diarrhoea. Nowadays it is more common to catalogue toxins depending on how they work, although the older classification is still of use. Luckily toxins fall into just three main groups, so it is necessary only to grasp three general concepts. Each type of toxin can be likened to a different type of military campaign. Some toxins attack like a battalion of tanks, some (like diphtheria) hit behind the enemy front line to disable supply lines, while others still are more devious and attack vital communication systems to cause confusion and disarray.

Although diphtheria toxin is the best understood of all toxins, it was not the first for which the mechanism was solved at the molecular level. That distinction belongs to a toxin made by *Clostridium perfringens* (originally called *C. welchii*), a bacterium notorious as the cause of gas gangrene. This disease often occurred in warfare, leading to terrible tissue damage and it was particularly bad in the trench warfare of the First World War. The bacterium was present in the soil and contaminated wounds, which often could not be cleansed for a long time because of the nature of the warfare. In 1941, Marjorie MacFarlane and colleagues at the Lister Institute in London found that the *C. perfringens* toxin attacked the membrane of cells. Clearly this is a very different type of mechanism from that of diphtheria toxin.

The *C. perfringens* toxin exemplifies one group of toxins that target the membrane of the cell. At the simplest level, the membrane of a cell is the container that prevents the contents from escaping—rather like a balloon that holds the helium inside. It defines the outline of the cell, but it does much more than that. Nutrients to keep the cell alive have to pass across the cell membrane. The membrane is a thin film that is hydrophilic on its two faces and hydrophobic inside, and is designed to prevent molecules passing through. Special mechanisms are incorporated into membranes to enable the cell to choose which molecules can pass in or out. The membrane is also an electrical insulator that enables the cell to maintain a voltage difference from the outside. This is essential for various reasons, the most obvious being in nerve cells that use electrical signals to transmit their messages. Crucially the membrane is the cell's interface with the outside world. Chemical signals released elsewhere in the body arrive at the membrane and are then processed and interpreted by the cell. The integrity of the membrane is essential for normal cell function and it is therefore no surprise that its disruption is a serious challenge to the cell. It is also not surprising that toxins have targeted this vital cellular feature.

The *C. perfringens* toxin is an enzyme called phospholipase that degrades some of the lipids (hydrophobic fats) in the membrane. This enzyme attacks only phospholipids.* Several bacteria make similar toxins. These include the bacteria found in some cheeses, *Listeria*, named after Joseph Lister, and *Pseudomonas aeruginosa*, which is one of the main bacterial complications in cystic fibrosis. The discovery of the activity displayed by this toxin was another big advance in toxin understanding—the proof that a toxin could be an enzyme. This fact largely explained the extreme potency of these molecules. Not only did toxins attack key targets, but each molecule was capable of modifying many molecules of the target.

Several bacteria make enzymes that attack membrane proteins. Many of these damage specific proteins—one example being *Porphyromonas gingivalis*,† a bacterium that is found in the mouth and that

---

*A fat molecule that contains a phosphate. Phosphate is an assembly of four oxygen atoms and one phosphorus atom. Phosphates are the way that nature uses phosphorus for many purposes in the cell.

†*P. gingivalis* is an anaerobic bacterium that does not grow in the presence of oxygen. Like many such bacteria, *P. gingivalis* has a foul smell, and working with it does not make you popular with colleagues!

has been linked to gum disease. Here the bony attachments between the jaw and the teeth disappear, loosening the fixation of the tooth to the jaw. This is a common disease that affects about 10 per cent of the adult population to some extent. In the worst cases the teeth will fall out. The *P. gingivalis* protease attacks specific proteins in the membranes of the target cells. These proteins sit in the membrane to allow the cell to sense its neighbours and pass signals into the cell, to reassure it that it is happily beside similar cells. By interfering with these proteins, *P. gingivalis* drastically affects normal functioning of the cell. Another interesting toxin is made by *Bacteroides fragilis*, a bacterium that is commonly found in the gut. Its toxin cleaves E-cadherin, a membrane protein of critical importance for cell–cell interactions. As with *P. gingivalis*, degradation of the target protein significantly perturbs signalling processes—in this case causing the affected cell to grow abnormally.

Many toxins damage the cell membrane through an entirely different mechanism, by organising several of their molecules into a ring shape and inserting this into the membrane to form a hole, or pore, in it. These toxins were discovered in the early 1980s and there are hundreds of them.

At first sight, pore-forming toxins might seem to be less sophisticated than the toxins that we have met so far. After all, there is no doubt that diphtheria toxin is very clever, because it displays several activities. Similarly, the membrane-attacking toxins are not just enzymes: they have to escape from the bacterium, and then survive in a potentially hostile environment before they meet the target cell. However, we should not be fooled into thinking that the pore-forming toxins are poor cousins within the family of toxins, just because they do not have enzymatic activity. Once released from the bacterium, each toxin molecule has to interact with other toxin molecules to organise themselves into an oligomeric* ring comprising several molecules. This ring of toxin molecules has to display hydrophobic patches on its outside so that it can breach the hydrophilic and hydrophobic features of the membrane to insert into it.[1] However,

---

*The suffix -mer is scientific shorthand for something with a number of units in it. A monomer is a single unit, a dimer, trimer and tetramer have two, three and four units respectively. A polymer has many units, while an oligomer has several—undefined but not as many as a polymer.

in the time between leaving the bacterium and inserting into a membrane, the toxin monomers, and perhaps the oligomers, have to stay dissolved in the fluid bathing the cell. This implies that they have a hydrophilic character, otherwise the hydrophobic patches would interact and the toxin molecules would clump together. In the same way that the diphtheria toxin uses local conditions to signal when to unmask its hydrophobic character, it is believed that the pore-forming toxins must do the same, although the process is not understood in such detail. Not only that, these toxins have to move into the membrane and stop after insertion without going right through into the cell. All this information is coded into the toxin molecule—nothing else helps it to insert into the membrane.

Many of these toxins are called haemolytic toxins. This means that they can split open, or lyse, red blood cells. As these are easy cells to obtain, they have been commonly used for the analysis of toxin action. When lysed, the red haemoglobin spills out. Just because a toxin can lyse red blood cells under laboratory conditions does not necessarily mean that this is what happens naturally. Moreover, use of these cells is rather an unfair test of toxin action, because a red blood cell is unable to fight back. This cell, the only one of all cells in our body, does not have a nucleus and thus does not have any genes. It is therefore unable to mount a response to the attack by switching on genes that could be used to repair the damage.

Real nucleated cells that can express new genes can respond. They try to repair the hole by putting a patch of membrane over it, similar to a self-repairing bicycle tyre. So the battle between these toxins and their target cell is more complicated than was first thought. Nevertheless these toxins are still very effective against normal nucleated cells. The induced loss of electrical potential across the membrane may be an important aspect of their function, but at least some of these toxins may have other functions. Some pore-forming toxins may move inside the cell to attack intracellular targets. The adenylate cyclase toxin made by *Bordetella pertussis** is an enzyme that works inside the cell, but part of it looks like a pore-forming toxin. The pore-forming ability enables it to cross the cell membrane. The VacA toxin from the cancer-causing bacterium, *Helicobacter pylori*, makes pores, not just

---

*Bordetella pertussis causes whooping cough.

in the cellular membrane, but also in some internal membranes, leading to the formation of large vacuoles (or membrane bags) inside the cell. It also causes a variety of other effects in ways that are not understood at present.

A considerable number of bacteria that are important in human infections make this type of toxin. These include some of the bacterial *Clostridium* species, *Escherichia coli*, and *Listeria*, *Staphylococcus* and *Streptococcus* species.

The membrane-damaging toxins, particularly the pore formers, launch a frontal bombardment on the integrity of the cell. However, the other two classes of toxin display a more subtle form of attack. The great majority of these toxins interfere with the ability of the cell to interpret signals that arrive at its surface.

The ability of any cell to sense its outside environment is crucial. Communication is even more important in a multicellular life form. Humans have about 10 million million cells that have to act in a co-ordinated manner. These cells have to be instructed how to behave and/or to tell other cells how to behave. The chemical messengers that a cell receives can instruct it to grow and divide to make two new cells, or they might order it to change into a specialised cell with a particular function.[2] In addition, cells have to know that they are with neighbours, that is in the correct place. For example, liver cells need to be reassured that they have liver cells as neighbours, skin cells want to be beside other skin cells, and so on.

The significance of these processes is emphasised if we consider what happens if a cell acquires the ability to grow without receiving either chemical or neighbourly signals. If first of all a cell changes so that it does not need to receive the reassuring signals released from other cells, it will grow and divide to form a little colony of dysfunctional cells. This in itself is usually not too dangerous. The growing colony of cells can be left to grow slowly or if necessary removed by the surgeon's knife. If the misbehaving cells begin to break down neighbouring cells, to spread through and past them, this invasive property is more worrying. The cells must be removed. Moreover, some of these dysfunctional cells sometimes then acquire the ability to grow somewhere in the body where they should not be able to grow, that is beside cells that are not like them. Lots of colonies of these rebellious cells can then form. These dispersed cells that are out of control will seriously disturb the normal function of the body. Not only does this abnormal growth physically block normal processes; these wayward cells often release chemical signals that perturb the careful balance in the body. This spread of these dysfunctional cells is called metastasis and is very dangerous. Removal of the primary growth will not cure the problem and a way has to be found to kill off these multiple groups of wayward cells. This is difficult because they retain many of the properties of the original cells from which they were derived. As they are generally growing faster than normal cells, their DNA can be more easily damaged than that of normal cells.[3] Radiation (radiotherapy) or specific chemicals (chemotherapy) can be used to try selectively to kill the rogue cells.

The previous paragraph has described the process of cancer and the ways in which it can be combated. The concept of cancer as abnormal growth was understood centuries ago, and operations were performed without anaesthetics in desperate attempts to cure sufferers. This perception of cancer as a disease of abnormal cell growth developed around the same time as the concept of the gene in the late nineteenth and early twentieth century. It was realised that there must be some genetic damage in cancer, although it was unclear what it could be, and many apparently conflicting ideas were put forward. Outside influences, such as radiation and some chemicals,[4] were known to lead to cancer. It was not until the identification of the genetic material and its copying mechanism by the early 1960s that the stage was set for determining the molecular basis of cancer. However, there was no consensus then about which genes were changed. Although some therapies against cancer were successful without a full understanding of exactly how they worked, many scientists believed that more targeted therapies would become available only when the changes that occurred in cancer were better appreciated. That breakthrough began in the late 1970s and early 1980s.

The first real clue came from analysis of the chicken sarcoma virus, a virus that had been shown in 1911 by Peyton Rous to cause a particular chicken cancer. Much later, when the individual genes of the virus were isolated, it was found that one alone, called *src*, had the effect of transforming cells (making them cancerous). *Src* makes a protein known as a tyrosine kinase, an enzyme that puts a phosphate group on to the amino acid tyrosine in other proteins. This discovery

. was a huge step forward and generated great excitement. It fitted with the view that many had taken that cancer might be an infectious disease, and suggested that some viruses might contain oncogenes, genes that cause cancer. The discovery shortly afterwards that this gene was in perfectly healthy animals and humans was shocking. As these animals did not have cancer, the gene in these circumstances was labelled a proto-oncogene. Shortly afterwards another oncogene was discovered, called the *ras* gene—it was also present in healthy animals and humans. It was not a tyrosine kinase, but it could also transform cells. At the time, these two genes sat in little unconnected islands of knowledge, and it was not clear why our cells contained genes that potentially could cause us so much harm.

This little detour brings us back to signalling, because it turned out that *src* and *ras* are each signalling proteins. They do not sit in isolation in the cell—they are linked into pathways of signalling proteins that transmit and interpret signals arriving at the cell surface. They are not there to cause us harm; their normal role is to receive signals that activate them and then pass the signal on in a relay of information. An obvious consequence of this role is that to work properly these signalling proteins have to remain off until switched on by an incoming signal: after passing the signal down the line they must then return to the off mode. It was discovered that the transforming *src* and *ras* genes were mutant versions of normal *src* and *ras*. The mutations changed the molecules so that they were permanently switched on, and transmitted their signals continuously without any requirement for an incoming signal. As these are the sorts of signal that tell the cell to grow, it is not surprising that mutations in these genes can lead to cancer. Mutations in *ras* are found in about 30 per cent of human cancers (in some types more often than in others) and, although *src* mutations are relatively rare in human cancers, greatly increased activity of the protein is found in many cancers.

Of course it is not as simple as that. A normal cell has to acquire several mutations before it can free itself from its normal regulation and grow independently of incoming signals and without being beside similar neighbours. A further complex set of proteins regulates DNA synthesis and the decision of the cell to divide. The process whereby DNA replication is followed by cell division leading to two new cells from one old one is called the cell cycle. It is regulated by a set of proteins that were identified primarily by work by the British scientists Paul Nurse and Tim Hunt, for which they were awarded the 2001 Nobel Prize for Physiology and Medicine, along with the American Leland Hartwell, who had earlier identified and defined genes responsible for cell cycle control.

Not surprisingly, some of these cell cycle proteins are found to be mutated in human cancers. There are various inbuilt safeguards: the cell can fight back by deciding to kill itself,* when it finds that something is seriously wrong with its regulation of growth or if its genes have been changed. Mutations in the proteins that regulate cell suicide are also extremely common in cancers. Thus cancer turns out to be a disease caused by mutations that disturb the normal functions of signalling proteins, cell cycle proteins, and proteins that regulate the self-killing programme. These are closely linked processes. None of the proteins involved is there to cause cancer and all are crucial to the routine operation of the cell. The importance of these molecules to normal cell function makes them wonderful targets for clever molecules such as toxins.

Signalling proteins are built into highly interconnected pathways of different components. The first component of a pathway is located at the cell membrane (where the cell interacts with its environment) and often the end of a pathway triggers the activation of particular genes. Given that the cell is very complex, with an organisation that has been honed by millions of years of evolution, the multitude of interconnecting signalling pathways at first seems too difficult to understand, but there are really only a few, relatively simple, basic principles. The complexity arises from the number of pathways. The basic idea of a signal integration process is easy to understand, whereby different types of signal are interpreted and analysed to produce a limited number of outcomes. An analogy would be the different signals that we might receive which would determine whether we decide to stay in, go out, or even commit suicide. A complete absence of signals (no friends, no phone calls, text messages or emails, no input from television, etc.) might induce suicide. Messages from friends might encourage us to go out, but information about the state of the

---

*Cells are generally programmed to kill themselves if they do not receive reassuring signals telling them what to do. This suicide programme is called apoptosis.

weather, or about our lack of cash, would tend to have the opposite effect. It is an everyday event to process this sort of information and decide on an appropriate outcome.

Nature has adopted a relatively limited repertoire of signalling modules from which to build its signalling networks. The phosphate ion is common to most of these. This highly charged chemical group comprises a phosphate atom surrounded by charged oxygen atoms and is attached to the amino acids in proteins by enzymes called protein kinases. It is later taken off by enzymes called protein phosphatases. One group of kinases phosphorylates (that is, attaches phosphates) to either serine or threonine amino acids, whereas another group, including Src, attaches phosphate to a tyrosine amino acid. The addition of a phosphate group changes the behaviour of the protein. Kinases are a major class of signalling protein, many of which are activated by phosphorylation via a different kinase. The human genome has over 500 different protein kinases and 150 phosphatases—all thought to be involved in signalling, either to switch on a signal or to dampen it down. One of the anthrax toxins attacks a kinase.

Some kinases sit across the membrane and become activated when a signalling molecule passing the outside of the cell binds to the part of the molecule that projects out into the cell's environment. This interaction leads to activation of the kinase activity inside the cell. Other kinases are found as intermediaries in signalling pathways, and some activate proteins that bind to the genes in the nucleus to turn them on.

Another major class of signalling molecule comprise the G-proteins, named because a molecule of guanosine triphosphate (GTP) or guanosine diphosphate (GDP)* is bound to them. In its resting state a G-protein is bound to GDP, but when activated this is exchanged for GTP. The G-proteins do not modify their targets; instead the activated form binds to other proteins to change their activity. The signal is deactivated by the G-protein itself, which cuts one phosphate from GTP to give GDP and the G-protein returns to its resting state. One type of G-protein sits on the inner side of the membrane and receives signals directly from receptors called G-protein-coupled receptors

---

*Guanosine is one of the four bases or nucleosides that make up the nucleic acids DNA and RNA. GTP has three phosphates, whereas GDP has two.

(GPCRs for short), which criss-cross the membrane. The GPCRs are the largest category of proteins in the cell, which is one measure of their importance. Many drugs bind to GPCRs and so act through G-proteins. There are four classes[5] of these G-proteins, each of which activates different types of effectors in the cell. Several bacterial toxins act on the G-proteins.

Other G-proteins, called small G-proteins, operate in a similar way, but occur in the middle of signalling pathways, and are important in practically every aspect of cellular function—growth, movement, shape, differentiation, and not surprisingly cancer. It turned out that Ras was one of these. Indeed this group of about 50 proteins is often described as the Ras superfamily of small G-proteins. Many toxins attack these signalling molecules, in particular the Rho subfamily.

Some special forms of the nucleotides* act directly in signalling. The first signalling molecule ever identified was cyclic AMP,† so called because the phosphate is joined to the rest of the molecule in two places to form a circle. Cyclic AMP is made by the enzyme called adenylate cyclase, and broken down by an enzyme called phosphodiesterase. Cyclic AMP and its sister molecule, cyclic GMP, each bind to proteins to activate them. Some toxins interfere with the normal regulation of these cyclic nucleotides. There are a few other signalling modules, but none of these is known to be a target for bacterial toxins.

Although there are many examples of toxins that attack the membrane, there appear to be only a few examples of the group of bacterial toxins that bind to cell surface receptors and mimic normal messengers arriving at the cell surface. The toxin STa (for [heat] stable toxin) is one of the many toxins produced by that well-known bacterium *E. coli*.‡ STa binds to and activates a receptor on cells that line the small intestine. This receptor is normally activated by a hormone called guanylin, but STa locks onto the receptor to activate it chronically. Part of the guanylin receptor is inside the cell and is an enzyme that makes the small signalling molecule cyclic GMP. STa causes artificially high levels of cyclic GMP which activate special

---

*The building blocks for DNA and RNA, each made up of a nucleoside, a sugar, and phosphate.   †Adenosine cyclic 3':5'-monophosphate.
  ‡Even within one bacterial species, such as *E. coli*, there is specialisation. Different types of *E. coli* cause different diseases and express different toxins.

pores in the membrane, via signalling pathways that are not yet fully worked out. This in turn causes transport of liquid out of the cell into the gut. To the person whose gut is being so insulted this results in a bout of traveller's diarrhoea.

An intriguing group of toxins called superantigens was found to act at the cell surface to trick the immune system. The immune system normally recognises a foreign bacterium by digesting its proteins into small fragments, and the 'molecular signature' of these fragments, known as an antigen, is remembered so that it can be recognised and dealt with if the bacterium is encountered again. This is a very specific process. The fragments are held in a small cleft in a protein, called MHC (major histocompatibility complex), on the surface of one cell. Immune system cells called T cells each have a receptor, part of which is unique to each T cell and is pre-programmed to recognise a unique fragment. If a T cell meets its matching peptide in the right type of MHC molecule, of which there are several, the T cell is activated, makes more copies of itself, sends out chemical signals, and is primed ready to combat the pathogen if it meets it again. As a result of the precision of the recognition process only a very small percentage of the T cells are stimulated at any one time.

A superantigen bypasses this delicate and precise system by binding to part of the T-cell receptor that is common to many T cells, so activating a lot of immune cells inappropriately. Although a normal antigen might activate around a millionth of the T cells, a superantigen can activate around five per cent of them. The immune reaction produced is complete overkill and the primed cells do not even recognise the pathogen. Vast amounts of chemical signals are released, which call other cells, the neutrophils, to the fight. The neutrophils stream into the blood system and can clog up the fine capillaries, the narrow blood vessels in our organs that deliver the nutrients and oxygen from the blood. The result is shock and major organ failure can ensue.

Superantigens are produced by some strains of *Staphylococcus aureus* (best known in its highly antibiotic-resistant form, MRSA*) and *Streptococcus pyogenes* (a cause of many different human diseases, from sore throats to kidney infections and scarlet fever). Toxic shock

---

*MRSA stands for methicillin-resistant *Staphylococcus aureus*.

syndrome toxin (TSST), a toxin released by *S. aureus*, hit the headlines in the 1980s when it was associated with the incorrect use of
tampons. Bacteria, which were naturally present on the body, were
able to grow to high numbers in the rich blood meal that was being
supplied at body temperature.

A very different type of surface-acting toxin is not a protein, but a
lipid. This toxin is called endotoxin and is a component of the bacterial cell wall, called lipopolysaccharide* (LPS). LPS is found only in
those bacteria that do not take up the Gram stain, the Gram-negative
bacteria. Endotoxin can be boiled and still retain its activity. It binds
to a surface receptor[6] that triggers signalling mechanisms which, similar to superantigens, lead to shock. LPS is responsible for hundreds
of thousands of deaths even in the developed world through diseases caused by *E. coli*, *Salmonella* species and other Gram-negative
bacteria.

The story of endotoxin began back in the 1890s and initially
reflected the antagonism between Louis Pasteur and Robert Koch. It
began with Richard Pfeiffer[7] who had joined Koch's team in Berlin
in the mid-1880s. Koch had by this time discovered cholera and
Pfeiffer was instructed to work on this disease to try to identify its
toxins, as Koch viewed (correctly) that the disease had the hallmark
of one linked to toxin production. Pfeiffer's experiments in 1892
produced unexpected results that did not fit with the picture produced
by diphtheria toxin. First, it appeared that bacteria killed by specific
antiserum were toxic when injected into animals. Second, and amazingly, even bacteria killed by exposure to 100°C were highly toxic and
retained about a tenth of their starting toxicity, suggesting that the
toxin had unusual properties and was attached to the bacterium.[8]

This heat-stable toxin was found in many different bacteria. The
term 'endotoxin' was chosen to reflect Pfeiffer's view, wrongly as
it turned out, that the toxin was inside the cell. He also thought
that endotoxin did not stimulate an immune response. Alexandre
Besredka, a Russian recruited to the Institut Pasteur by Ilya Metchnikoff, showed that an immune response against the toxin could be
induced under certain circumstances, for which he was later honoured as a member of the Imperial German Academy of Scientists—

---

*This means made of lipid and saccharides (sugars).

further proof of the improving relations between the Pastorian and Kochian camps. It was, however, 1933 before it was discovered, by André Boivin at the Institut Pasteur, that endotoxin was not a protein, and a further 30 years before its chemical structure was understood in detail.

The most famous and notorious toxins are the final group those that act inside cells. These follow the pattern set by diphtheria toxin. They are all AB toxins, where a B domain (or region) deals with binding and uptake of the toxin and the A domain is an enzyme that modifies a key function in the cell. In some of these toxins, the B component is a separate protein that is not chemically joined to the A component. Practically all these intracellularly acting toxins can be discussed in just two broad themes: toxins that prevent proteins being made and those that attack signalling molecules. Many of these targets are GTP-binding proteins (G-proteins),* which suggests that bacterial evolution discovered long ago that these proteins were important, and designed a method for attacking them.

Exotoxin A from *Pseudomonas aeruginosa* behaves exactly like diphtheria toxin to modify EF-2 and inhibit protein synthesis. The cause of dysentery, *Shigella dysenteriae*, makes Shiga toxin which also inhibits protein synthesis. It was initially assumed that it worked like diphtheria toxin, but in the 1980s it was shown that it attacked the RNA scaffold of the ribosome to hinder the binding of another factor necessary for protein synthesis. Shiga toxin is also found in *E. coli* O157, the cause of hamburger disease. This bacterium became notorious for the 1996 outbreak of disease in Scotland that infected around 500 people, killing 21 of them. In fact, it caused disease worldwide, including the USA, where it is called hamburger disease, and also Japan, where 10,000 people were affected in 1996. A major scandal occurred in Canada where the water supply in the small town of Walkerton became contaminated with this bacterium, resulting in the death of up to 14 people. *E. coli* O157 is carried harmlessly by cattle, but has now collected a gene that enables it to make a very dangerous toxin and in humans it can infect the urinary tract to produce the dangerous haemolytic uraemic syndrome (HUS). The plant

---

*The target of diphtheria toxin, EF-2, is also a GTP-binding protein, so that the major theme of attacking G-proteins applies here too.

toxin ricin, notorious as a biological weapon, has the same enzymatic activity as Shiga toxin.

The ADP-ribosylation of a G-protein, the method of attack used by diphtheria toxin, is employed by several other toxins, but is adapted to hit different targets. Michael Gill showed that cholera toxin ADP-ribosylates G_s, one of the signalling G-proteins at the membrane, leading to its chronic activation. This in turn causes an increase in the small signalling molecule cyclic AMP. As cyclic AMP regulates transfer of salts and water from the body into the gut, its wildly elevated production in cholera leads to massive passage of fluid into the gut and the terrible watery diarrhoea of cholera.

Compared with the remarkably short time that elapsed between the discovery of *Corynebacterium diphtheriae* and the proof that it produced a toxin, it was over 70 years between the discovery of *Vibrio cholerae* and the discovery of its toxin independently by S.N. De and N.K. Dutta in India in 1959. A further 10 years later, cholera toxin was produced. Still later it was shown by Jan Holmgren that the toxin was a different version of the AB formula: the A part was separate and sat on top of a pentameric* doughnut of B proteins.

Pertussis toxin, produced by *Bordetella pertussis*, the bacterium that causes whooping cough (indeed called pertussis in the USA), causes ADP-ribosylation and inactivation of a different G-protein, G_i. As activation of this G-protein normally inhibits cyclic AMP production, its modification by pertussis toxin also leads to an increase in cyclic AMP. As whooping cough is a lung disease, the effects occur there. It is still not clear how pertussis toxin helps the bacterium, but it probably prevents the immune system in the lungs from working properly, so enabling the bacterium to establish there and cause disease. Whooping cough is a very dangerous disease in early childhood. Although the numbers of deaths per year are decreasing, the statistics of the World Health Organization (WHO) show that around a quarter of a million children die each year of this disease. Immunisation is effective at reducing this carnage. In older childhood the disease is still unpleasant and complications can arise because of secondary invaders—pathogens that attack the body after it has been weakened by one pathogen.

---

*Pentameric means that it has five subunits.

The other types of G-proteins, the small G-proteins, have been subjected to a considerable onslaught by bacterial toxins, particularly a subfamily of them called the Rho proteins—probably because these proteins are so important for normal cell function. Some of these toxins cause ADP-ribosylation of their target—indeed that was how the Rho family was identified. It has been shown in the last 10 years that other types of toxins also attack Rho. Some bacteria from *Clostridium* species make very large toxins that inactivate Rho by attaching a sugar molecule to it. These bacteria often live harmlessly and under control in our intestines, but can grow when antibiotics kill off harmless bacteria and can then induce antibiotic-associated diarrhoea. A related toxin was responsible for the unexplained deaths of 35 heroin addicts in the UK and Ireland, in the summer of 2000, who were injecting heroin that was infected. Other toxins activate the Rho by modifying one of its amino acids so that it is unable to cut GTP down to GDP to switch itself off, and remains locked in its active form. This mechanism was worked out in the laboratories of Patrice Boquet in France and Klaus Aktories in Germany. One of the toxins from *B. pertussis* and yet another toxin from *E. coli** carry out this reaction.

The role of these toxins in disease has not been established, but we can safely assume that they are not up to any good and very probably prevent the immune system from working normally. The *E. coli* toxin that activates Rho is mainly found in bacteria in urinary tract infections; *E. coli* is the main cause of these infections. The activation of Rho raises levels of an enzyme called cyclo-oxygenase-2 (COX-2), which elicits other concerns. COX-2 is an enzyme involved in many processes and is the target for many painkillers, such as aspirin, that reduce its activity. People on long-term aspirin therapy have lower rates of certain cancers, which is thought to correlate with the lower activity of COX-2. So increased activity of COX-2, particularly over a long period of time, may be a risk factor in cancer.

A long time before signalling inside cells was appreciated, it was known that some toxins affected a very different type of signalling. The deadly effects of the neurotoxins had been recognised but not understood for centuries. The disease of tetanus was particularly unusual, leading to Hippocrates' description mentioned earlier. Botulism

---

*The *E. coli* toxin that activates Rho is called cytotoxic necrotizing factor (CNF).

was identified more recently, probably because the floppy outcome of botulism was less noticeable than the tightly clenched muscles in tetanus.

Work on tetanus began in earnest in 1894, when Antonio Carle and Giorgio Rattone in Italy showed that the disease could be transmitted: pus taken from someone who had died of tetanus could reproduce the disease in rabbits. Arthur Nicholaier identified the bacteria responsible in the same year and Shibasaburo Kitasato later grew them in pure culture.

The name botulinum comes from the Latin for sausage (*botulus*) because of its close association with the consumption of sausages. The bacterium was named by Müller in 1870, but some very carefully thought-out work in the 1820s had already laid out a far-reaching view of botulism poisoning. The authorities in Stuttgart had been concerned, in the early nineteenth century, with the great increase in food poisoning and issued a warning in 1802 about the consumption of blood sausage. The University of Tübingen was asked to help, and they gathered reports from local doctors. One of these doctors was a local medical officer called Justinius Kerner.

Kerner produced a detailed description of the disease symptoms, and then began animal experiments using extracts from bad sausages that were likely to be contaminated. He even experimented on himself. He correctly deduced that the sausages contained a potent poison that developed in bad sausages deprived of air. Furthermore, it worked only on the nerves that controlled the muscles, not on the sensory nerves or the brain. Based on the information that he had obtained, he even suggested using the poison in small doses to block nerve function in various diseases. Over 150 years would pass before these far-sighted ideas were developed further (see Chapter 8). His was a considerable accomplishment, achieved before any understanding of the role of bacteria, let alone toxins, in disease.

By the mid-twentieth century it was known that the botulinum toxin was the most deadly of all toxins, as only a few billionths of a gram were needed to kill a human. The disease is usually not caused by an infection, but the consumption of food contaminated with toxin previously produced by the bacteria. Arnold Burgen (later Sir Arnold Burgen FRS) and colleagues identified that the botulinum toxin inhibited signal transmission at the junction between the nerve and muscles by blocking the release of the neurotransmitter chemical from nerves, although it was not clear how this occurred. Normally the released neurotransmitter bound to a receptor on the muscle fibres to pass on the signal.

The mystery of how tetanus and botulinum could prevent nerves from functioning was finally solved in 1992. It was shown that these toxins were each proteases, that is enzymes that cut up other proteins. This discovery came from the newly identified protein sequence of the toxin. This sequence contained a motif or signature often found in proteases. The neurotoxins do not chew up any protein at random. They are highly specific for some of the proteins involved in nerve transmission. The release of the neurotransmitter is a highly regulated process. A membrane-enclosed vesicle containing the chemical is held inside the cell and is moved to the membrane when the nerve is stimulated to fire. The vesicle interacts with the membrane to release its contents—rather like a bubble rising to the surface of a liquid and emptying its gassy contents to the atmosphere, except that it is more precisely controlled. The docking mechanism for the vesicle relies on several proteins, three of which can be cut by botulinum toxins in a precise manner to prevent the docking process. There are seven slightly different types of botulinum toxin, each of which cuts one of the proteins involved in the docking process.

Tetanus cuts the same protein as one of the botulinum toxins. Why then does tetanus produce spastic paralysis where all the muscles clamp tightly, whereas botulinum toxin causes the opposite effect, producing a flaccid paralysis where there is no muscle tone? The answer is that these two toxins attack different types of nerve cell. That choice is decided by the B domain of the toxin—the part that causes binding and uptake. Tetanus toxin travels along nerves to attack nerves in the central nervous system, whereas botulinum toxin blocks signals peripherally where the nerves are trying to make muscles contract. The ability to target the powerful function of these toxins is now being exploited for novel uses.

Despite the pivotal role that anthrax played in the birth of microbiology, it was the mid-1950s before it was decisively proved to be a toxin-based disease, in pioneering work carried out by Harry Smith and his colleagues at Porton Down. There are two anthrax toxins. They are slightly unusual AB toxins, because three factors are involved.

We could call them A_2B toxins, as the protein that binds to the cell, a protein called protective antigen,[9] can then bind to either of the two toxic factors to carry them into the cell. Very recently, one of the anthrax toxins, lethal factor, was shown to be a protease by Stephen Leppla in the USA, and also around the same time by a collaboration between Cesare Montecucco's group in Italy and Michèle Mock's group in France. Lethal factor attacks a crucial signalling protein that is activated by the Ras protein, although it is not clear how this helps the bacterium or causes disease.

The other anthrax toxin also impinges on signalling. Instead of modifying a signalling protein, Leppla showed in the 1980s that this toxin actually makes one, the small molecule cyclic AMP—the molecule that is also increased indirectly by the actions of cholera and pertussis toxins. This toxin, the adenylate cyclase toxin, is known as the 'edema factor', because it causes 'edema' (American spelling for oedema), or swelling, in animals. Evolution had a particular problem with this toxin. As most toxins are designed to work on a target found only in a human or animal, but not in bacteria, there is no danger that the toxin will cause its producer any harm. However, cyclic AMP is used both by our cells and by bacteria. It turned out that the adenylate cyclase toxin works only when it is bound to another protein— one that is found exclusively in animal and human cells, but not in bacteria.

It is rather sad that anthrax has become well known for all the wrong reasons—not as a disease understood and conquered, a shining triumph of nineteenth and twentieth century science and medicine, but as a potential terrorist threat.

Bordetella bacteria also make an adenylate cyclase toxin and, similar to the anthrax toxin, it is activated only inside our cells. However, it is not an AB toxin. As discussed earlier, part of the *Bordetella* adenylate cyclase toxin behaves like a pore-forming toxin, so that it can insert into the membrane in order to pull the enzymatic part across it into the cell. So *B. pertussis* makes three protein toxins, all of which affect signalling processes: two alter the concentration of cyclic AMP and one activates the Rho proteins. Bordetellas have one further weapon: part of its cell wall sloughs off and this, the tracheal toxin, affects the trachea or windpipe to inhibit the normal removal of unwanted particles from the lungs. This is carried out by the beating of tiny hair-like structures, called cilia, on the cells that line the trachea. Tracheal toxin prevents the cilia from beating. *Bordetella* attacks the cell with a cocktail of toxins.

One group of toxins has recently been found to attack an even more important target than the signalling molecules. These toxins attack at the very heart of the cell, to damage its DNA. The cytolethal distending toxins (CDTs), so called because they are lethal and usually cause elongation of cells, are produced by many different bacteria[10] and were first described about 15 years ago. They appeared to be very similar to enzymes known to degrade DNA, but then it was shown that CDT moves into the nucleus where it directly cuts DNA. Cells treated with CDT behave as if they are suffering from radiation damage, leading to the possibility that exposure to this toxin could lead to cancer.

Why do bacteria make toxins? In some cases this is obvious. The cholera toxin induces the excretion of vast quantities of diarrhoea infected with the bacterium. This is an excellent strategy for transmitting the bacterium and infecting new people. Diphtheria toxin damages the tissues in the throat, thereby generating a supply of food for the growth of the bacterium, and this is probably a tactic of use to many disease-causing bacteria; the related consequence, that the toxin kills the host, may be incidental to the bacterium. Many toxins appear to target the immune system, and this is obviously a good strategy to adopt. In other cases it is not yet clear what advantage the toxin confers to its host bacterium, although there must be some advantage because natural selection would long ago have eliminated any gene that was not of use.

A factor common to all these toxins is target selection. In every case toxins have selected targets of the utmost importance to the cell that is under attack, such as membrane integrity, protein synthesis, signalling processes, or DNA itself. It is hardly surprising that so many toxin diseases are so dangerous.

6

WHY ARE PLAGUE AND TYPHOID SO DEADLY?

A further layer of cunning

I am fond of pointing out to students that nine-tenths of the cells that make each of us what we are happen to be bacterial—only 10 per cent are our own cells. Of course this is not quite as horrifying as it seems, because the bacteria that inhabit our gut and live on our skin are about 10,000th the size of our own cells. Therefore, although our bacteria outnumber our cells by 10:1, they constitute only about 0.1 per cent of us in terms of weight. Nevertheless this highlights that not all bacteria are bad for us. Not only do we live quite happily with our harmless bacterial companions, but we actually need them to help us to digest food and to ward off dangerous bacteria.

The ability to provoke damage to our cells is the key attribute of pathogenic bacteria that distinguishes them from harmless bacteria. Toxins clearly fulfil this destructive function extremely efficiently. However, one group of bacteria behaved as if they made toxins, although none could be identified, and it was unclear how they caused disease. This group includes bacteria well known through their constant appearance in the news. Most notorious was *Yersinia pestis* which causes the plague. The *Salmonella* bacteria are also infamous: *S. typhi* is the agent of typhoid, and *S. typhimurium*[1] and *S. enteritidis* lead to food poisoning salmonellosis. *Shigella* bacteria, an important cause of dysentery, also fall into this group. These bacteria puzzled scientists for a long time and many unsuccessful attempts were made to look for toxins. *S. typhimurium* was subject to particularly intense investigation, with some reports of the discovery of a *S. typhimurium* toxin. However, work on these reported toxins could not be reliably repeated by others.

The answer to this conundrum turned out to be utterly unexpected, and led to the discovery of a new group of toxins that had a novel way of getting into cells. It had been known from about 1950 that *Yersinia* bacteria grown in the laboratory showed a strange effect that relied on the presence of calcium ions. The bacteria refused to grow when calcium salts were removed from their nutrient broth, and this in some way correlated with their ability to cause disease. Then, in 1990, it was discovered in Guy Cornelis's laboratory in Belgium that the removal of calcium corresponded to the release of several proteins from the bacteria. A great deal of work over the next 15 years mainly from two groups—of Guy Cornelis and of Hans Wolf-Watz in Umeå, Sweden—changed the story from an enigma into a new insight that can now explain what happens in some detail.

First, it was shown that HeLa cells[7] died when *Yersinia* bacteria were attached to them. A *Yersinia* protein called YopE* was essential for this outcome, but disappointingly the addition of crude preparations of *Yersinia* bacteria to the cells had no effect. Injection of *Yersinia* proteins directly into the HeLa cells, using very thin glass tubes, killed the cells. Convinced that these proteins were important for the appearance of virulence, both groups looked for the YopE protein in cells infected with *Yersinia*. This led to the startling discovery that the bacterial protein appeared inside cells during infections. The bacteria had injected the proteins directly into the target cell. Proteins isolated from the bacterial cells had been completely inactive because, unlike conventional toxins, these toxic proteins have no intrinsic mechanism for penetrating the host cell. These bacterially injected toxins are commonly referred to as effector proteins.

The bacterium itself controls entry of these toxins into the host cell. It produces a complex injection apparatus, called the injectisome, that penetrates the two membranes of the bacterium and the membrane of the targeted host cell. This injectisome contains a central pore through which the injected proteins pass to gain entry to the cytoplasm. This pore remains plugged until it is needed. When the injectisome has penetrated a host cell membrane, the proteins

*Yop stands for *Yersinia* outer protein.

that are to be injected are moved through the opened pore to attack the target cell. *Attack* is probably not strong enough; *take over* would be more accurate, because these toxins, even more than their conventional cousins, take control of the host cell and regulate it for their own devious purposes. This method of protein export (secretion) by bacteria through an injection tube is referred to as either type 3 or type 4 secretion, because these are the third and fourth methods discovered whereby bacteria deliberately export proteins across their membranes.

The injection machinery is yet another demonstration that bacteria are not simple creatures, because the injection engine comprises about 30 different proteins. The proteins in the secretion machine are very similar in all bacteria that have this system, enabling biologists to search for the tell-tale signs of this mechanism in other bacteria. The genes for the injectisome are usually clustered together on a special region of the genome called a pathogenicity island. The discovery of these regions was at about the same time as the discovery of type 3 secretion. Pathogenicity islands are large stretches of DNA that are usually long enough to code for about 20–40 genes. They can be recognised by the unusual bias of their nucleotide composition, because they have a larger than expected number of two of the bases that make up their DNA compared with the other two. As the over-represented bases are adenine and thymine, these segments of DNA are often called 'AT rich'. One explanation for this difference from the other genes is that the pathogenicity island has arrived in the genome of the bacterium from an outside source in the relatively recent past, several million years ago. These regions code for a cluster of genes that carry out linked functions, so that the arrival of a cartridge of 'dangerous genes' in the genome of a bacterium can convert a harmless bacterium into a pathogenic one. There are many examples of genomes being changed suddenly by this type of mechanism, perhaps brought in by a bacterial virus, and pathogenicity islands often show the remnants of a viral signature.

Yersinia bacteria use type 3 secretion to inject six different effector proteins into its target cell, the macrophage—the immune cell that is usually the first line of defence against foreign invaders. The bacterium is highly selective in regulating which proteins can pass into the targeted cell, and only the protein effectors designed to work in the target cell move through the port. It is not clear at the moment how this gating works, but it appears to be quite complicated, because there does not appear to be a common pattern of amino acids that serves as a signal. This is a highly sophisticated system.

The *Yersinia* toxic proteins really are weapons of mass destruction: five of them interfere with the cell's signalling processes. As with conventional intracellular toxins, the Rho signalling proteins are key targets. YopE, the protein that led to the concept of type 3 secretion, behaves as if it is a normal signalling protein that inactivates Rho. It does not act as an enzyme to modify Rho. This was a surprising result, because all conventional toxins that act inside cells have so far been enzymes. Another injected protein is a protease that cuts Rho, so that it is no longer attached to the inside of the cell membrane, and thus becomes inactive.

Yet another *Yersinia* protein is a kinase that adds phosphates to proteins to alter their activity, whereas a different protein has the opposite effect—it is a powerful phosphatase that removes phosphates. The outcome of the action of these two proteins is to block uptake of the *Yersinia* bacteria, and to downregulate the immune response.

Less is known about the other two injected *Yersinia* proteins. One binds to cellular protein kinases, to block their phosphorylation and activation, and thus prevent activation of two key signalling proteins. It has also been shown to induce the death of macrophages. The final protein is even more of a mystery, although its presence is essential for disease. It is known to move into the nucleus of the cell, where we can safely assume that it is up to no good.

The calcium part of this story was that the protein that plugged the injection port was partly held in place by calcium ions. When the calcium ions were artificially removed from the growth medium, the proteins that should have normally been kept in the bacterium for injection into a cell were released harmlessly from the bacterium into the growth medium.

The terrifying onslaught that *Yersinia* launches on its victim does not completely explain why the plague wrought such havoc over the centuries. Other factors are important in trying to understand why it was so deadly.

It is unclear who first described plague symptoms because of the confusion between diseases caused by other infections and true plague. Some think that Rufus, a doctor from Ephesus, a town now in Turkey, described the plague around AD 100. However, it could be argued that he was describing a different disease from the vagueness of his description.

Many people consider that the first description that ties in with the symptoms of true plague are those of the plague of Justinian in 540. This is important because it now appears that the plague is a relatively new disease. The complete DNA sequence of *Y. pestis*, obtained in 2000, showed that *Y. pestis* was remarkably similar to another yersinia bacterium, *Y. pseudotuberculosis*, which causes a relatively mild gastro-intestinal disease—a very different disease from plague, in terms of both its severity and its disease profile. When *Y. pestis* became distinct from *Y. pseudotuberculosis*, it lost the function of a large number of genes. Many of these are still present but with so many changes that they do not work. The presence of these genes, called pseudogenes, suggests that *Y. pestis* is in a state of flux, having recently found a limited lifestyle where it does not need such a wide variety of different genes for different environmental conditions. Analysis of these sequences suggests that *Y. pestis* emerged as a new species distinct from *Y. pseudotuberculosis* somewhere between 500 and 20,000 years ago. As the plague of Justinian was 1,500 years ago, this estimate can be revised to between 1,500 and 20,000 years ago. This is an incredibly short time in evolutionary terms.[3] So it is possible that the plague of AD 540 was the first time that plague was visited on the human population to such a devastating degree, and this might in part explain its high virulence and ability to cause such devastation.

Diseases often appear to be most virulent and dangerous when they first appear. However, it is not necessarily to a pathogen's advantage to be so highly virulent that it kills off all its available host. On its first appearance, a virus or bacterium has often just acquired some property that enabled it to cross into a new species, that is us. In this circumstance it can be highly virulent and often over time it becomes less so, preferring to live more in harmony with us. Several diseases show these characteristics.[4]

In addition, a new disease, or one that has suddenly re-emerged, will be more virulent because none of the population has any inbuilt immunity. The sudden appearance of plague in 540 and the 1340s may have hit a world population that had had no exposure to plague and was thus highly sensitive. As a disease develops, survivors of one round of disease are likely to have immunity when the next round appears, and at the same time anybody who displays natural resistance to the disease will prosper, producing resistant offspring and increasing the proportion of resistant individuals in the next generation. This is of course natural selection.

Several other factors could have compounded the severity of these plagues. In AD 535 a catastrophic world event appears to have occurred that led to a severe and prolonged drop in temperature—perhaps a period of 15 years of the lowest temperatures in 2,000 years. This is based on evidence gathered by Mike Baillie at Queen's University in Belfast using tree-ring widths in both Europe and the Americas, so it was clearly a global phenomenon. This ties in with reports of summer snow in China and of a dust veil that lasted 18 months. There are a couple of theories about this, each as controversial as each other. The first is that the Earth received a cosmic impact or our planet passed through a substantial meteor shower. The other is that a vast volcanic explosion took place—in particular, an explosion of Krakatoa is suggested. Whatever the explanation, it is clear that such an event would lead to widespread crop failure, famine, and general population disturbance.

The early fourteenth century also saw a succession of several years of particularly bad weather, leading to crop failures and associated malnutrition, both in Europe and elsewhere. This was aggravated by the large increase in population that had occurred in the years up to the start of the fourteenth century. It is believed that the onset of plague in China was preceded by several years of drought, followed by floods. In addition, France and England had begun what was to be the Hundred Years War just 10 years before the Black Death.* So in each case the global weather changes and other factors led to famine that was bound to have promoted increased susceptibility to infectious disease.

The life cycle of the plague would also have contributed to its severity. Plague is really a rat disease. There are several stages: first the *Y. pestis* bacteria grow in the gut of the rat flea, *Xenopsylla cheopis*.

---

*The intervention of plague forced a six-year truce from 1349 to 1355.

One of the genes that *Y. pestis* acquired in its split from *Y. pseudo-tuberculosis* is crucial for both its survival in fleas and its transmission to an animal host. This gene codes for the *Yersinia* mouse toxin—wrongly named as it turns out. It is an enzyme that attacks membranes and works best at 26°C, which is the temperature of the flea. This protein protects the bacteria from attack inside the flea and in many cases leads to a blockage in the flea gut. The starving 'blocked' flea still tries to feed, but this does not do it any good. So, when it bites a new host, it literally vomits its *Yersinia*-infected meal into it, so transmitting the infection. It is estimated that tens of thousands of bacteria are injected by each flea. This method of delivery straight into the bloodstream is a highly effective route, giving a rapid onset that overwhelms the immune system, allowing little chance for the body to fight back, and is surely one reason for the high virulence of plague. When rats are not available the flea attacks humans. It was noticed in both the plague of Justinian and the Great Plague that the disease started at the coast, which fits with the idea that it was brought in by rat-infested ships. Presumably the same pattern was observed for the Black Death.

The bacteria delivered by the flea into the blood get into macrophages that are our first line of defence. Although macrophages are adapted to engulfing and killing the bacteria, the *Yersinia* bacteria have other ideas. They appear to live happily in the macrophage, while adapting from life in a flea to life in a warmer mammal. However, bacteria that are taken up by neutrophils, which have similar functions to macrophages, are killed. As the patient experiences severe fever, the infected macrophages gather in the lymph nodes and, with the massive recruitment of phagocytes, the lymph nodes swell up hugely. Bubonic plague is so named because of these characteristic buboes, which are the grossly enlarged lymph nodes (up to a small apple size is often quoted) visible in the groin, armpit, and neck. They comprise *Yersinia* bacteria and dead and damaged cells. The bacteria, which have now adapted to a new warmer lifestyle, escape from the macrophages to live freely in the blood. They ungratefully kill the macrophages that they have been living in, and any other macrophages too. Any infection that directly attacks the immune system is on to a winning strategy. The bacteria can then multiply freely throughout the body. Delirium and lack of coordination soon follow, probably related to the high fever and the buboes become agonisingly painful. The infection rages through the body and death follows several days after the first signs. The net result is that bubonic plague kills between 30 and 90 per cent of those infected, although it does not spread from person to person.

If the bacteria establish in the lungs before the person dies, pneumonic plague has begun. Pneumonic plague begins abruptly with fever and often with a headache. Another sign mentioned in some reports was the sweet taste that victims noted in their mouths before the onset of fever. When plague changes to pneumonic plague, not only is it even more dangerous, reportedly 100 per cent fatal, but it can be spread by droplets from the inevitable blood-tinged coughing by the infected person. Not very much is known about the microbiology of pneumonic plague, but it is a rapid killer—death often occurs within two days of the first symptoms. Presumably, the bloody destruction of the lungs causes sufficient damage to lead to death.

Total ignorance of the cause or transmission of the disease added to its ability to spread. The Venetian quarantine of ships did not stop infected rats from escaping to spread disease. None of the other suggested remedies had any scientific basis. At the time of the Great Plague cats and dogs were thought to transmit the disease. About 40,000 dogs and 200,000 cats were slaughtered in London, although of course they might well have helped to limit the disease by killing infected rats.* The village of Eyam closed itself off from the outside world to prevent infecting the rest of Derbyshire after a bale of infected cloth from London's Great Plague transmitted the plague there. In doing so, Eyam lost over 70 per cent of its inhabitants. This tragedy is all the greater when it is realised that many more would probably have survived if the villagers had taken to the hills and avoided close contact with each other.

However, the cause of both the plague of Justinian and the Black Death still remains highly controversial. Graham Twigg has suggested that anthrax caused the Black Death. A recent and rather angry book by Susan Scott and Christopher Duncan puts forward the view that neither pandemic was caused by *Y. pestis*, and instead by a virus

*However, Daniel Defoe does refer to attempts to kill mice and rats, especially rats.

similar to Ebola.[5] Their views were based on the disease symptoms, what is known about the rat population at that time, and the incubation period and speed with which the disease spread. If the main disease had been pneumonic plague, however, it could have spread with that degree of rapidity and, as discussed above, the disease could have behaved differently at that period. They also suggest that the infection should have been less prevalent during the winter months, although during the Great Plague in London the disease did decrease during winter, reaching its peak in the summer. Although *Y. pestis* DNA was identified in the dental pulp of a victim found in a mass grave in Montpellier, France from a later plague, Scott and Duncan dismiss the importance of these results. Although they acknowledge that plague was around then, they suggest that it was not the most important disease.

This is an issue that will probably never be satisfactorily resolved. The finding of *Y. pestis* DNA in samples taken from old burial grounds does not prove that this was the cause of the pandemic, although similarly a failure to find such evidence would not prove that *Yersinia* was not involved. As the weight of evidence still supports the view that *Y. pestis* was the main culprit, we will take that as the best working hypothesis—although other diseases ongoing at the same time might have compounded and exacerbated these epidemics.

Although it is evident that there is still more to be discovered about how *Yersinia* takes over its victim's body, it is obvious that it launches a brutal attack on the host's defence systems almost before the host realises that it is under attack. As well as the vicious cocktail of injected proteins that are so sneakily delivered, some other strains of *Yersinia* make the conventional toxins STa (stable toxin) and CNF (cytotoxic necrotizing factor). The host has no chance.

The story of injected toxins is still new and is rapidly developing, as a growing number of bacteria are being shown to attack us using toxins delivered by injectisomes. All these bacteria are Gram negative, that is bacteria that do not hold on to the stain of Christian Gram, and thus are bounded by two membranes. How did bacteria come to have such a complex system? Not surprisingly it evolved from something else—that's how evolution works—making mistakes, that is mutations, and then selecting ones that give an advantage. The type 3 secretion machine is very similar to one that bacteria use for movement, in particular to move up chemical gradients. They do this by rotating a long wand, called a flagellum.* The arrangement of proteins in the rotating machine and in the injection apparatus is very similar. The type 4 secretion apparatus evolved from a different system—one linked to bacterial sex, whereby one bacterium injects DNA into another bacterium through a fibre called the sex pilus.

Currently much more is known about type 3 than type 4 secretion systems. As well as in *Yersinia*, these have been analysed in some detail in shigellae and salmonellae. Probably the best-known example of a type 4 system is found in *Helicobacter pylori*, the bacterium that causes stomach cancer. It injects a protein called CagA, which interferes with several signalling mechanisms; one of these mimics metastasis, the ability of cancer cells to spread to other sites in the body and thus the most important feature of carcinogenesis. Very recently, it has been shown that bacteria of *Bartonella* species, which can produce tumour-like effects in cells that can, however, be treated with antibiotics, use type 4 secretion to deliver effector proteins of unknown function.

Salmonella bacteria are an interesting group. Typhoid has been around a long time, and is endemic at times of war and famine. Before it was more or less eradicated in the west by improved water supply and sanitation, it was a constant cause of disease and death. Two of Louis Pasteur's children were lost to its clutches, and through the years typhoid has taken its toll of the famous and poor alike. The Emperor Augustus is reputed to have had the disease and recovered, although it could of course have been some other type of fever. Mozart is also believed to have recovered from typhoid. The earliest description of typhoid is generally attributed to the London doctor, Thomas Willis, writing in 1659. The Frenchman Pierre Louis, in 1825, described which internal organs were affected, but did not distinguish between typhoid and typhus. That was clarified by the American, William Wood Gerhard, about 12 years later. The bacterium itself was identified in 1880 by Carl Joseph Eberth, one of Robert Koch's disciples.

Typhoid is spread by contaminated food and water that has been fouled by infected faeces, as already mentioned in Chapter 2. The

---

*From the Latin 'to beat'—the weird cult of the flagellants was discussed in Chapter 1.

typhoid bacteria escape from the gut of an infected person and spread through the blood to attack organs such as the liver and spleen. This leads to high fever, headache, lethargy, and upset to gut function—either constipation or diarrhoea. Severe complications can lead to gut perforations and loss of blood. Delirium and shock are also sometimes reported.

Typhoid remains a major killer in south-east Asia. Worldwide the disease is estimated to infect at least 16 million people annually, killing 600,000. As a result of under-reporting the number may be far higher. A worrying trend is the appearance of antibiotic resistance. Old-style vaccines have limited efficacy and produce side effects, but new genetically engineered vaccines are currently being tested.

Although the disease remains a matter of great concern in Asia, complacency in the so-called developed world is certainly not appropriate. There was a substantial outbreak (469 cases) of typhoid in the city of Aberdeen in Scotland in 1964. This was traced to a contaminated tin of corned beef. When the tin's contents had been put through the meat slicer in a supermarket, the bacteria had spread to other products. A quick, and some have said overzealous, response by the local Medical Officer of Health, Ian MacQueen, and widespread media coverage led to the rapid containment of the outbreak—almost a self-imposed quarantine—and the all clear was sounded within 28 days. Unfortunately the newsworthiness of the story led to an unfair stigma being attached to the name of Aberdeen for some time after.*

A particularly interesting, but disturbing, source of disease is caused by the carrier state that Koch and others discovered in the early 1900s. Some people who have suffered typhoid, but who have recovered and show no symptoms of disease carry the bacteria that continue to grow slowly in them. The carrier state occurs in around two to five per cent of people who have had typhoid. These carriers still excrete bacteria and can therefore infect others. This is of particular concern with food handlers, as in the sad story of 'Typhoid Mary', who lived in the USA in the early twentieth century. Mary Mallon was an itinerant cook, originally from Ireland, who in a period of ten years moved between eight households bringing typhoid to seven of

---

*The Queen paid a visit to Aberdeen shortly after this time to show that it was safe to go there.

them. Scores of people were infected and several died. It was while she was cooking for a family in the then, as now, affluent Oyster Bay in fashionable Long Island, off New York, that her deadly role as a typhoid carrier came to light. Six out of eleven of the family on holiday came down with typhoid in 1906. The owners of the house were worried, because typhoid was by this time not a common disease in places such as Long Island. There was no evidence of contaminated water or food. A sanitary engineer, George Soper, was employed to investigate the outbreak and he noticed that it coincided with the time that Mary Mallon had been cooking for the family. She had moved on to a family in Manhattan when Soper found her. He brusquely approached her and accused her of spreading typhoid, demanding samples of her faeces and urine. She naturally rejected these demands.

Soper then traced her previous positions as cook and found that her employment had coincided with the appearance of typhoid elsewhere. Armed with this knowledge he arranged for her to be forcibly taken to hospital by order of the New York Health Department. She was found to be a typhoid carrier and placed in an isolation hospital on North Brother Island. She was allowed to leave after three years on the condition that she gave up cooking for others. Unfortunately she broke this agreement and was found cooking in Sloane Maternity Hospital in Manhattan. In her three months there, 25 staff had had typhoid and two had died. She was forcibly returned to her one-room cottage in the isolation hospital on North Brother Island for the 23 years until her death. She was the first typhoid carrier to be identified in the USA, although clearly there were others identified around the same time who were not treated in the same harsh manner. Her name has become demonised, even being used to name a heavy metal band.

The carrier state is also important because of the effect that this may have on the person who carries the bacteria. It is becoming clear that carriers have a higher risk of some types of cancer, particularly of the bile duct.* The original suggestion came from an analysis of typhoid carriers from a small outbreak of typhoid in New York in 1922. This study was published in 1979, and showed that carriers were more prone to these cancers. The later analysis of the Aberdeen

*The bile duct is linked to the liver and the gut.

outbreak showed that people who had suffered acute disease, but who had not become carriers, had no increased incidence of any cancer. This group was compared with a list of people known to be carriers, and it was found that the latter had a significantly increased risk not only of cancer of the bile duct, but also of pancreatic cancer—both cancers that are difficult to treat effectively. As a result of the nature of cancer, this does not mean that every typhoid carrier will get cancer, in the same way that not every cigarette smoker succumbs to lung cancer. It is just one piece of the jigsaw. However, given the numbers of people who contract typhoid annually, it is potentially a highly significant cause of disease and death. On a more positive note, the realisation that *S. typhi* can be linked to cancer may help our understanding of the role of other bacteria in cancer and also our understanding of cancer in general.

The absence of toxin production by any of the salmonellas was a puzzle for many years. The food-poisoning bacteria, although not in the same league of malevolence as *S. typhi*, still cause unpleasant and dangerous disease. The diarrhoea induced by *S. typhimurium* or *S. enteritidis* can be of violent and sudden onset, with the unfortunate victim passing blood from the damaged gut. Some of the food-poisoning salmonellas behave like *S. typhi* in spreading dangerously to sites around the body. *S. dublin* and the rare *S. choleraesuis* are particularly notorious for this. So, for these bacteria, as well as for typhoid, it appeared to be incongruous that they did not produce some type of toxin molecule that could explain the havoc that they caused.

The identification of several pathogenicity islands that coded for proteins involved in type 3 secretion in *S. typhi* and in other salmonellas went a long way to explaining how this group of bacteria could be so damaging. *Salmonella* bacteria are loaded with toxic weapons. There are two secretion engines: one is involved at the gastrointestinal stage of the disease and manipulates the attacked cell to force it to take up the bacteria and one is involved when the bacteria have penetrated beyond the gut into deeper tissues.

Salmonellas make a substantial number of effectors, many of which have similar activities to the *Yersinia* effector proteins and which attack Rho and other signalling proteins. As with *Yersinia*, *Salmonella* bacteria inject proteins that kill macrophages, and also some of unknown function move into the nucleus. There is clearly more to learn about *Salmonella* species, but at least why these bacteria are so dangerous is not now such a mystery.

The other bacterium with a type 3 secretion system that has been analysed in some detail is *Shigella*. This bacterium was first isolated by Kiyoshi Shiga in 1898 and causes dysentery—a nasty intestinal diarrhoeal disease spread by contaminated food and water which can lead to the excretion of bloody diarrhoea. The disease has been known about from early recorded history and claimed George Washington as one of its victims. Current estimates suggest that in excess of 150 million people are infected each year, leading to at least 600,000 deaths—some estimates are far higher. There are currently no vaccines, but these are being actively developed. As with so many bacterial infections, antibiotic treatment is effective against shigellosis, but antibiotic-resistant strains are now appearing.

Much of the analysis of the proteins injected by these bacteria has been carried out by Philippe Sansonetti and his colleagues at the Institut Pasteur in Paris. Rather like *Salmonella,* the injected proteins help the shigellas to enter cells. However, less is known about how these proteins work compared with *Yersinia* or *Salmonella.* One *Shigella* protein activates Rho proteins in an as yet unknown manner, whereas another appears to have the opposite effect. As with both *Yersinia* and *Salmonella, Shigella.* makes proteins that kill macrophages and some that enter the nucleus.

Thus, bacteria use these cunning secretion mechanisms to manipulate host cells in subtle ways and to knock out immune function. The mechanism of delivery is efficient, in that effectors are sent directly into the targeted cell. Although the translocation machine is complex, once designed and built by evolution it can be used for any suitable effector protein. The bacterium does not have to evolve separate mechanisms either to export a toxin through the bacterial membrane(s), or to get it across the target cell membrane. These toxins are perhaps the smartest of them all.

7

DEVIANT BIOLOGY

Weapons, espionage, and man's innate inhumanity

Biological weapons hit the headlines in October 2001 with the US anthrax attacks. The anthrax scare in the USA was followed by the discovery of ricin-making equipment in London in early 2003 and the topic has stayed in the news ever since. The possibility that Iraq had retained a biological weapons programme was the justification for the 2003 Gulf War and led to post-war repercussions when bio-weapons could not be found. There is therefore no doubt about the potency of these weapons to cause alarm—even without their wide-spread deployment.

Nuclear, chemical, and biological weapons have come to be known as weapons of mass destruction (WMDs). This overused phrase is rather an artificial classification, because conventional munitions can also be weapons of mass destruction, and indeed have been respons-ible for millions of deaths over the centuries. The trench warfare of the First World War and the intense conventional bombing of German cities towards the end of the Second World War each caused mass destruction.

The moral standpoint against chemical and biological weapons, and the illegal construction of nuclear weapons, is largely a political one that reflects the perspective and customs of the time. Europe and the USA have each used chemical weapons, and made and stockpiled biological weapons, whilst the only country known to have used a nuclear weapon is the USA.

Historically, the move from hand-to-hand fighting to more distant attack using guns began to remove the romantic view of fighting as a heroic undertaking where the braver and (usually) stronger man won, but where each side was willing to put themselves personally at risk. Technology based on human ingenuity became an important issue. Any fighter with vastly superior weapons is usually able to deploy them with limited risk to themselves, whether this is combat between clubs and bare fists, guns and swords, or nuclear attack against a non-nuclear state. The side with the poorer technology usually accuses the side with more advanced technology of cheating—that is until they catch up. The result of each increase in technology is that warfare becomes more dangerous, more brutal, and more indiscriminate. How these changes are perceived depends very strongly on viewpoint. The English admired the superior firepower of the English longbow over the crossbow at the battle of Crecy, and the use of fire ships when the English defeated the Spanish Armada is nowadays viewed as a brilliant piece of naval tactics, at least in England. Computer-guided Cruise missiles are often referred to favourably as 'smart bombs'.

Weapons of mass destruction are viewed as repugnant, particularly because they are seen to be highly indiscriminate in their attack. Of all the WMDs, bioweapons currently stir up the greatest fear. There are several reasons for this: first is our general lack of experience of these weapons and the fear of the unknown. The First World War was a chemical one and the Second World War became nuclear. The dreadful thought is that the next one, or even a smaller-scale conflict, could become biological. As the most dangerous agents are self-replicating, a small amount of an infectious agent could in the worst circumstances replicate out of control. This is of course exactly what happened naturally in the vast disease pandemics that have ravaged humanity. This could potentially lead to the ultimate human tragedy, the annihilation of the human race. Moreover, unlike the other two WMDs, the production of biological weapons does not require highly sophisticated technical equipment, and they have been described as the 'poor man's weapons of mass destruction'. In addition, the technology related to bioweapons is so-called dual use. A factory making vaccines can be changed to one making weapons the following week, making it difficult to discover and police weapon construction.

In addition, although the deployment of chemical or nuclear weapons is usually obvious,* a biological agent may not be discovered

---

*Chemical weapons generally act within minutes or hours.

for some time—indeed it may not ever be detected as a deliberate act,* and this can lead to further scare-mongering. It has been suggested by some that recent outbreaks of disease in animals and humans might not be natural, and might have a more sinister origin. This belief includes the 2000–2001 outbreak of foot and mouth disease in the UK, and even the SARS (severe acute respiratory syndrome) outbreak that began in China in early 2003. There is no evidence to support these suggestions.

Biological warfare is an ancient endeavour and not just a new threat that began in October 2001. Even when infectious disease was not understood, aspects of it were sufficiently appreciated for use in warfare. There are reports of arrows being dipped in manure or dead bodies over 2000 years ago. Hannibal in 184 BC had pots filled with snakes that were thrown on to the ships of the King of Pergammon. The bemusement of the sailors as the earthenware pots were hurled at them soon turned to terror as the pots shattered to release a writhing mass of poisonous snakes. Contamination of an enemy's water supply with diseased rotting corpses is also an ancient and effective tactic. It was later used during the American Civil War, when soldiers deliberately shot animals and left them to rot in ponds. This tactic is still being employed.† The device of catapulting plague-infected bodies into a besieged city may have led to the devastating spread of the Black Death. A similar attack was carried out by the Russians against the Swedes, at Ravel in Estonia, in 1710.

In 1763, British officers suggested giving smallpox-infected blankets to Native Americans, and the smallpox epidemic in the Ohio Valley Indians has been attributed to this. The British may also have used smallpox during the American War of Independence. However, devastating as these episodes in biological warfare may have been at the time, each of them was more or less a spur-of-the moment attack, rather than a long-term military tactic.

It was during the First World War that a systematic strategy was begun to develop bioweapons in a scientific manner. One driver for

---

*Although the Rajneeshees' act of *Salmonella* contamination in 1984 affected 751 people, it was a year before it was realised that it was not just a natural cluster of food-poisoning cases (see page 130).

†The Nazis poisoned reservoirs with sewage in 1945, and the Yugoslavs poisoned wells in Kosovo in 1998.

this change was the concern that the other side might launch a pre-emptive bioattack, and there was also a desire to find a weapon that could break the stalemate and slaughter of the trenches. The extent of the weapons' programmes that were initiated remains far from clear. However, 1915 is generally thought to be the start of biowarfare directed at animals. A German, Erich von Steinmetz, entered the USA (dressed as a woman) with the intention of poisoning horses with glanders, a bacterial disease caused by *Burkholderia mallei*. This bacterium has also been considered for use against humans. Unfortunately for him, but luckily for the Americans, the cultures had not survived the journey, as he found when he had them checked at a microbiological laboratory. Anton Dilger was another German who attempted to attack horses with glanders and anthrax, although his attempts were also unsuccessful.

Towards the end of the First World War, the USA considered using the plant toxin ricin as a weapon. Ricin is made by the castor oil plant, *Ricinus communis*, which has been cultivated since ancient times for the oil in its seed. Castor oil was widely used during both World Wars when there was a shortage of other oils—this is the oil that the Castrol Company first used and took its name from. It is also used as a laxative. The plant grows easily in warm climates, so the raw material is readily available. The pulp remaining after the oil has been extracted from the seeds with solvents contains between two and five per cent ricin, which can easily be purified.[1] As a million tons of castor beans are processed annually, there is the potential to make a lot of ricin.

It was suggested that ricin could be either attached to bullets to increase the chance of inducing fatality or used as a volatile cloud for an inhalational attack. Code-named compound W, ricin was later developed during the Second World War by the USA and the British, to the extent that ricin bombs were prepared and tested. There are no reports that they were ever deployed.

Attempts to legislate against bioweapons internationally have been on-going since 1874, but with limited success. The Peace Conference at The Hague banned poisons in 1899, but did not stop the chemical attacks of the First World War. In 1925, the Treaty of Geneva was drawn up to ban the use of biological weapons. It was never ratified by Japan, and only in 1975 by America. Nevertheless, by the 1930s many countries that had endorsed the Treaty were attempting to develop bioweapons. The Russians had been struck by the fact that typhus and other natural diseases had killed millions towards the end of the First World War and in its immediate aftermath. They set out to turn typhus into an effective bioweapon. The Germans set up the Military Bacteriological Institute in Berlin with the intention of developing anthrax as a weapon. In 1936, France and the UK each set up dedicated laboratories, the latter installed at Porton Down in rural Wiltshire. Canada also began looking at the possible use of anthrax and botulinum toxin and plague, whilst the USA began work at Camp Detrick in Maryland to prepare botulinum toxin, anthrax, *Brucella* and *Chlamydia psittaci*.* By 1943, this unit had around 4,000 personnel.

The laboratories that would have the greatest impact on the use of biological weapons were set up by the Japanese. In 1930, the ultra-nationalistic Major Shiro Ishii (Figure 7.1), having completed his medical studies and his PhD, returned to Japan from a two-year visit to Europe. He was the driving force behind the Japanese biological weapons programme over the next 15 years. Ishii has been described as a brilliant student, with a ruthless, self-centred, and arrogant streak. He was grovellingly subservient to his superiors, but treated his inferiors poorly. He also had a reputation as a womanising heavy drinker, with a specific interest in young girls.

The argument that Ishii advanced was that biological weapons must be very effective because the Geneva Convention had banned them. His military masters supported him strongly and he was given substantial resources to carry out his plans. He was also rapidly promoted, reaching the rank of Lieutenant General by the end of the war.

In the 1930s Japan was flexing its imperialistic muscles, particularly against its Chinese neighbours. Ishii began his biological weapons' work in Tokyo soon after his return, but needed somewhere to test his ideas. Japan had occupied Chinese Manchuria in 1932, and this appeared to be the ideal location. Ishii went there the same year, and the first weapons' laboratory was in the centre of Harbin city, with a more remote detachment for human experiments in the village of Beiyinhe. It was eventually abandoned in 1937, by which time his

---

*The causative agent of psittacosis, a bird disease that can transfer to humans. It is a generally debilitating disease that can have serious consequences.

7.1 Shiro Ishii.
(Photograph provided by Ignatius Ding of the Alliance for Preserving the Truth of Sino-Japanese War.)

most notorious unit at Pin Fan, Unit 731, was being built. Pin Fan was also close to Harbin city. This vast complex comprised about 150 buildings on a site of three square miles. A further large complex, Unit 100, was created in Changchun in the same year. In all, a total of 18 units were eventually set up across the areas of south-east Asia occupied by the Japanese. Many of these specialised in one particular aspect of biological or chemical warfare. Unit 731 investigated a long list of possible infections for weapons use, including plague, typhus, cholera, typhoid, anthrax, botulism, and smallpox.

The Japanese scientists developed gentle release mechanisms for those agents, because high explosive bombs could easily destroy the potentially fragile organisms. Porcelain containers could break with smaller explosives and release their deadly payload unharmed.* A mixture of anthrax spores and shrapnel was considered to be particularly effective. The unit experimented with the delivery of *Yersinia pestis*, the agent of plague, by releasing infected fleas. In some cases, contaminated food, such as chocolate, was distributed to infect the population.

This vast operation had several levels. Experiments were carried out under laboratory conditions at Unit 731 and elsewhere. These agents were then used against the indigenous Chinese population to test out their effectiveness in the field. The quantities of agents produced were huge. Finally they produced material for use on the battlefield and there is clear evidence that they were used in war.

Besides the experiments at Unit 731, Ishii's men also carried out trials of various bioweapons on the civilian Chinese population. This campaign began in 1938, when planes dropped contaminated balls of cotton wool in China. This method of delivery was particularly favoured—not surprisingly because it minimised the risk to the deploying forces. The same technique was later used to drop fleas mixed with grain into Manchuria to infect the local rat population, resulting in a local outbreak of plague. Other methods of delivery included feeding cakes contaminated with *Salmonella typhi* to Chinese prisoners of war, who were then released to initiate an epidemic of typhoid. In another episode they sprayed Chung Shang village with *Yersinia pestis* and afterwards went in to investigate how effectively

*Hannibal's earthenware pots filled with snakes illustrated the same principle.

they had delivered disease. Around 1,000 people perished in this and neighbouring villages. They torched the village to prevent further spread. Dog food contaminated with cholera was another method of release that could result in human disease.

These weapons were used in the battlefield against both the Russians and the Chinese. During the Chekiang campaign in 1942, the Japanese commanders are believed to have deployed biological weapons, including cholera, against the Chinese. Although vast numbers of Chinese were killed, it is estimated that about 10,000 Japanese became ill and 1,700 died. Certainly cholera is known to have killed at least 1,000 people in Guang Dong province in 1943.

After their defeat by the Russians at the battle of Nomonhan, the Japanese poisoned the water supply at the Mongolian border with bacteria, causing typhoid and dysentery. For various reasons, bio-weapons were not deployed against the Americans or British. However, towards the end of the war there were plans to send kamikaze planes loaded with plague-infected fleas to the USA's west coast. That plan was seemingly curtailed by the atomic bomb attacks that ended the war, and perhaps also by a belief that America's resolve would be strengthened and not weakened by such an attack.

What took place at Unit 731 and elsewhere in Japan was a shocking and grotesque breach of human rights.[2] The experiments conducted at Unit 731 were to a large extent carried out on humans. Chinese prisoners and civilians suffered most, although it is believed that captured Australian, American,[3] British, Russian, and other nationals were also used in barbaric infection experiments and other types of experiment of the utmost depravity. This was justified in the minds of the perpetrators as necessary to see what happened in a field situation, and the general catch-all that 'it was war'. The people used were referred to dismissively as 'marutas', meaning logs—part of the dehumanising process so often used by those who inflict torture. Infected people, including pregnant women, were tied down and in many cases dissected conscious, as it was suggested that anaesthetics would affect the behaviour of their vital organs. Babies were cruelly experimented on. There was not, and is not, the slightest real scientific justification for any of these deeds. It is instructive that one eyewitness recalled that live female dissections attracted a bigger audience than male dissections. The motive of many was clearly the perverted power experienced by those carrying out what amounted to extreme torture.

The numbers that died during this terror are unclear, but most estimates suggest at least 3,000 people died at Unit 731 from 1940 to 1945. More may have perished before these dates, and at the other sites. The remnants of the experiments, the tortured corpses, were disposed of in the camp crematorium. The true horror of what happened there will never be properly known, because much of Unit 731 and the other facilities was destroyed and the remaining prisoners murdered as the end of the war approached. It is alleged that infected fleas were released and caused deaths in the surrounding area in the years after the war. Canisters of chemical poisons left over by the Japanese military from this period still cause health problems.

Facts about the Japanese military's use of bioweapons are also hard to come by and difficult to verify, although estimates of the total numbers killed by Japan's bioterror are put at up to 250,000. Appalling as these crimes against humanity were, it is perhaps even worse that none of the main perpetrators was tried or punished for these war crimes, and many obtained advanced positions in post-war Japanese society. Ishii himself died of laryngeal cancer at the age of 69 in 1959. The American military struck a deal with the Japanese scientists that granted immunity in exchange for information about the experiments. It is suggested that they had two motives. First, they wanted to ensure that the Russians did not get hold of the details of the experiments. Second, they realised that the Japanese were well ahead in understanding and delivering biological weapons, and they hoped that the information gleaned at Unit 731 would help them to catch up. Two scientists from the US Weapons Unit at Fort Detrick, Dr Edwin Hill and Dr Joseph Victor, interviewed 22 of the Japanese scientists. They took cases of papers back to the USA, but whether this information was of any use is not known. Indeed these might have been the details that were returned without translation. It is not clear whether those who agreed to the amnesty realised that human experimentation had been carried out at the time of this agreement.

The Russians put 12 of the Japanese scientists on trial in the remote eastern city of Khabarovsk in December 1949. Most were sentenced to 20 years, but all appeared to have been allowed back to Japan in the mid-1950s for reasons that are not clear. It has been suggested that knowledge of the trials was suppressed in the west because of fears that the American amnesty would become known. This could have been especially explosive if news that Allied prisoners were tortured had come out.

The secret finally leaked out in the early 1980s. In 1984, a student came upon an old box of unsorted papers in a second-hand book-shop in Tokyo. These papers had belonged to a former officer and described in detail every stage of someone dying of tetanus, from the inception of the infection to its final conclusion. Such information could be known only as a result of deliberate human infection. Furthermore, human bones of non-Japanese origin were discovered in 1989 on a construction site in Tokyo where the Medical Military College had been located at the time of Ishii's first involvement—before Unit 731 was constructed. Since then several groups in the USA, China, and Japan have tried to raise awareness of the issue.

Although the numbers killed cannot be compared with those who perished in the Nazi death camps, the barbarity was certainly on an equal scale. Although the Germans did not overtly use bioweapons, they used them indirectly, by incarcerating people in squalid conditions, with little food, non-existent sanitation, stress, and exposure to constant infections. Rather chillingly, they let Nature help them with their extermination policy.

It has been suggested that the Soviets used tularaemia against the invading Germans at the battle of Stalingrad in 1942. Although many Germans may have perished, many Russians contracted the disease and died. This bad experience is thought to have led the Russians to regard bioweapons as too dangerous for the battlefield and only of use when deployed from afar.

During the Second World War, the British had experimented with anthrax. Some suggest that it was to be used only as a weapon of last resort. Operation Vegetarian aimed to kill animals and thus weaken the German food supply, as well as entering the food chain to kill civilians. The method of delivery was to be via cattle cake infected with anthrax spores, and was set for delivery by release from aircraft. The anthrax spores were produced at the Ministry of Agriculture Veterinary Laboratories in leafy Weybridge, now the Veterinary Laboratories Agency. Five million infected cattle cakes were produced and were ready to be deployed. However, as the 1944 invasion of the European mainland successfully turned the war, deployment of anthrax was shelved. Later all the infected cattle cake had to be incinerated.

The effectiveness of anthrax was tested on the small island of Gruinard just off the west coast of the Scottish mainland in 1942. Sheep were tethered in several places and small anthrax bombs located on gantries above the ground were set off. There was also some deployment from aircraft. All the animals died. Several sources report that the infected sheep were thrown from a cliff that was then dynamited to cover them. One sheep carcass escaped this burial process and turned up on the mainland, where a dog chewing on the carcass contracted anthrax. The dog survived, but passed the disease on to 25 sheep, which did not. The British government had to make up an elaborate story to cover its tracks.

Gruinard was off limits until its decontamination over 50 years later, because of concern about the great longevity of anthrax spores.[4] Spore samples prepared by Pasteur had been tested about 70 years later and found to be viable, as were some American soil samples kept for a similar time. In 1981, an environmental group called 'Dark Harvest' raised awareness of the Gruinard Island situation. They dumped a packet containing anthrax-contaminated soil near to the Porton Down complex, claiming that it had been taken from the island.* Surveys of the anthrax contamination on Gruinard had been carried out annually since 1948, but had not measured the degree of contamination. A more detailed study that was carried out in 1979, and published in 1981, showed that contamination was limited to certain areas, and importantly only within the top 10 cm of the soil. The government decided to decontaminate the island and 280 tons of formaldehyde† were applied to the infected areas. After further testing, sheep were allowed to graze on the island. None of them caught the disease and people were again allowed to visit.

It has been suggested that the British supplied botulinum toxin prepared at Porton Down in specially adapted grenades which were thrown at the hated Reinhard Heydrich, the originator of the 'Final Solution' for the extermination of the Jews, in Prague in 1941. He died

---

*In addition a box of soil was deposited at a Conservative Party meeting, but this turned out not to contain anthrax.

†Formaldehyde joins proteins together. It has been used in vaccine production since the early work on diphtheria and tetanus to inactivate toxins.

seven days later. Others who were also hit by grenade fragments survived, and it has not been proved or indeed disproved that Heydrich's infected wounds and blood poisoning were the result of contamination by botulinum toxin. An alternative possibility is that the infection came from dirt collected by the grenade fragments as they passed through the door of his car, or from his clothing.

After the Second World War and during the Cold War the USA bioweapons' programme expanded. *Francisella tularensis*, the cause of tularaemia, was made on a large scale. The main characteristic of this disease is its highly infectious nature. Fewer than 10 bacteria are needed for a lethal infection and it can be transmitted as an aerosol. *F. tularensis* lives inside cells, but little is known about how it causes disease. The USA also made typhus bacteria and anthrax. *Brucella suis* and staphylococcal enterotoxin B were developed as agents to incapacitate temporarily rather then kill an enemy. Bombs to deliver agents such as *Brucella* were developed by the early 1950s, and for tularaemia in the 1960s. The USA carried out several covert attempts to measure how well a bioweapon would disperse by releasing harmless 'simulants', for example *Bacillus globigii* (to simulate *Bacillus anthracis*) in public places. In 1950, San Francisco was sprayed with a simulant, and it is known that *B. globigii* was released in the New York subway system in 1966. These acts only became public knowledge many years later. At the same time, the USA carried out a defensive programme. This had benefits beyond bioweapons—for example, they were able to supply vaccines to Egypt when a natural outbreak of the viral disease Rift Valley Fever broke out there.

In 1969, President Richard Nixon decreed that the USA would no longer have a biological weapons' programme. By this time, there was public revulsion at such weapons, prompted by various events. The Dugway incident in 1968, when around 6,000 sheep died in the appropriately named Skull Valley area close to the army testing ground in Utah, is thought to have been caused by leakage of nerve gases. Dumping of surplus chemicals at sea also caused concern. Then, in July 1969 the army announced that 24 people had been contaminated by a leakage of the nerve gas sarin at the US base at Okinawa. This story was particularly damaging because it had not been previously known that such weapons were stockpiled there. By this time, the military view was that such weapons were of limited use and of dubious value, a view that had already led to a reduction in the programme. Furthermore, by banning the 'poor man's weapon of mass destruction', the USA could occupy a high moral position and yet retain a more effective weapon that was too costly to be afforded by lesser nations, that is nuclear weapons. A Biological and Toxins Weapon Convention of 1972, banning the production and stockpiling of bioweapons, was widely ratified, including by the then Soviet Union and the USA.

All the same, the Soviet programme continued. In 1979, reports came from Sverdlovsk, now renamed Yekaterinburg or Ekaterinburg, a small town in the Urals about 1,000 miles (1,600 km) from Moscow, of an outbreak of anthrax in which nearly 70 people had died with many more being hospitalised. The total death toll is now variously estimated between 200 and 1,000. The developing story led to a new chapter in the Cold War, with American spy satellite photographs suggesting the presence of decontamination trucks around Compound 19 at the site. This was where the accidental release had occurred. The Russians dismissed the story as wild American propaganda, claiming that the anthrax deaths were caused by eating contaminated meat. If this had been the case, cutaneous and gastric anthrax would have been the main disease, whereas most deaths were caused by pulmonary inhalational anthrax. A report by the renowned American biologist Matthew Meselson, based on the evidence available at the time, thought that the Russian explanation was plausible. It later turned out that he was wrong, and indeed the Americans had grossly underestimated the extent of the Russian programme.

In 1989, the story began to unravel. The Russian Vladamir Pasechnik defected while at a London conference. He confirmed the existence of Biopreparat—a vast network of laboratories making bioweapons out of various organisms. The defection three years later of the deputy chief at Biopreparat, Ken Alibek*—previously known by his Russian name of Kanatjan Alibekov—revealed the extent of the Soviet operation. The Soviet programme had made the plague bacterium, *Yersinia pestis*, resistant to antibiotics. In addition they had made a more potent version of *Francisella tularensis* and high-quality ricin.

---

*Ken Alibek wrote a book, *Biohazard*, about the Russian operation after his defection.

Finally, in 1992, Russia, through President Yeltsin, admitted that the Sverdlovski incident had been a biological weapons accident, although other Russian officials continue to deny the existence of any such programme. It apparently had taken five years to clean up the infected area at Sverdlovsk, because the hardy anthrax spores are resistant to normal methods of disinfection.

The Soviet bioweapons' programme had been scaled down during Mikael Gorbachev's reign. One factor that prevented Russia from abruptly closing Biopreparat was the perceived danger of throwing many highly trained microbiologists into unemployment. The worry was that these disillusioned scientists might be offered the temptation of lucrative employment by rogue states keen to develop their own weapons. What appears to be happening now with US acceptance and support is a gradual transfer of these factories to vaccine units.

The Iraqi weapons programme has caused the greatest recent interest. Following the first Gulf War in 1991, the UN set up UNSCOM (United Nations Special Commission) to investigate whether Iraq had made biological and chemical weapons. Working under difficult conditions, the inspectors found clear evidence for the production of a whole range of noxious substances, including ricin, botulinum toxin, anthrax, and the cancer-inducing aflatoxin. Research had been conducted at Salman Pak with a small team of scientists, while production took place at the Al Hakam factory. Bombs had been manufactured at the Muthanna plant. It is clear that delivery systems were also being developed.

South Africa is also believed to have engaged in biological weapons, but there is no evidence that they were ever used. Other countries, often referred to in the west as 'rogue states', have been accused of having bioweapons, including Iran, Syria, North Korea, and Libya.

All the above information is publicly available. However, we do not know if there are any other, as yet undisclosed, biological weapons being made. Given human ingenuity and natural aggressiveness, it is unlikely that this threat will ever go away. Many might consider that the use of such weapons in a total war situation is not very likely, because of the potential for serious worldwide consequences.[5] Internationally agreed and verifiable surveillance procedures are essential before people will be convinced that bioweapons are no longer being produced at the national level. However, it is undisputed that terrorist use that is either state sponsored or perpetrated by small individual groups is very much harder to police and control and that such use is to be expected. Indeed fanatics ordered by their leader to deploy such weapons may not even be concerned about their own safety.

Biological terrorist activities have mainly been carried out by fanatical cults linked to a charismatic religious leader. Although some of these have resulted in deaths and illness, luck has largely intervened to prevent a much worse outcome.[6]

RISE was a far-right racist group that was being established by two teenagers, Steven Pera and Allen Schwander, in Chicago in the early 1970s. They planned to contaminate the water supply with *typhi* and were within days of carrying out this act when they were caught. Both skipped bail, but Pera later surrendered and was sentenced.

Bhagwan Shree Rajneesh is perhaps best known for the fleet of Rolls Royces given to him by his devoted followers and the fact that the followers wore saffron clothes. He was an interesting character, who had started his commune in Poona in India in the 1960s. By the 1970s he was attracting a following of disaffected Europeans and North Americans who were looking for some sort of mystical enlightenment. The followers came from all walks of life and were often intelligent, well-educated people. As a badge of allegiance, they wore a picture of the master round their necks. They believed that the guru offered a new perception on the confusing world.

Not everyone was taken in by the Bhagwan's charm and, by the early 1980s, because of local opposition, the sect was keen to find a new home. They found one in Oregon, in America, with a property previously known as 'the Big Muddy Ranch'. The new settlement became a town called Rajneeshpuram. It had its own police force and was to a large extent integrated into the Oregon legal processes. Although there was a continual low level of grumbling opposition to the Rajneeshees' activities, things really came to a head in their clashes with the Wasco County Court, which was seeking to limit the continual requests by the sect to take over more land. The Court comprised three elected members, two of whom were opposed to the Rajneeshees. Ma Anand Sheela,[7] who was second in command at Rajneeshpuram, decided that decisive action was needed.

About 12 people were involved in the plot, of whom the most influential (and the most sinister) was known as Ma Anand Puja, a nurse who was in charge of the Rajneesh Health Center. Those who did not like Puja referred to her as 'Dr Mengele', after the evil Nazi doctor. She was a loner, and fascinated by poisons and death. She allegedly had talked about culturing *S. typhi*, the cause of typhoid, and HIV. However, the weapon they chose to use was *S. typhimurium*—a well-known cause of food poisoning.

Their first documented use of *S. typhimurium* was when three of the commissioners visited the Rajneeshees in the summer of 1984. The two who were opposed to the group were given water contaminated with the bacterium and both became sick afterwards, with one needing hospitalisation. Later that year, at the time of the local election, the Rajneeshees hatched a more ambitious plan to swing the vote towards their preferred candidates. As well as bringing in thousands of homeless people from around the country to vote for their chosen candidates, they decided to poison other voters in order to prevent them from voting. The group visited 10 salad bars in the Dalles spreading cultures of *S. typhimurium* on to the salads. A total of 751 people became ill, and a large number had to be hospitalised, although it took a year before the State authorities realised that these cases resulted from deliberate contamination. With the increased awareness of the possibility of this type of incident, it is unlikely nowadays that any such clustering of cases would not raise obvious suspicions. The cult also tried, but apparently failed, to contaminate the local water supply, either with *Salmonella* or with raw sewage. These incidents were not repeated and by the following year both Sheela and Puja had left the cult. Two cult members were eventually prosecuted.

The Japanese cult Aum Shinrikyô, led by their guru Asahara Shôkô, is known for its murderous sarin attack on the Tokyo underground that killed 12 people and affected around 4,000. However, it is less well known that they had earlier tried out biological weapons. The size and structure of their operation made the Rajneeshees look almost amateurish, and it is only by good luck that a serious number of fatalities were avoided. The cult numbered over 10,000 members, with assets of around £200 million. They had begun in the mid-1980s as a meditation group, but by the mid-1990s their aim was not a minor one—they wanted control of Japan and a place in world politics against the USA. It was clear that any means to achieve this was acceptable to them. Aum Shinrikyô was well organised, with structures that mirrored the Japanese government. Their 'Minister for Health and Welfare' was Endô Seicho, a molecular biologist who had worked at Kyoto University.

It is believed that they tried out botulinum toxin on several occasions. In 1990, and later, they drove trucks through Tokyo and other cities spraying botulinum toxin, although it is not known whether the toxin was active. An attack in 1995, when briefcase bombs were placed in the Tokyo underground, is thought to have failed because the person told to fill the bombs decided against it at the last moment. This failure led to the use, five days later on 20 March, of the nerve gas sarin, which instigated the investigation into their activities.

The cult also tried to use anthrax. They owned a building in Tokyo, which was turned into a production facility. Amazingly, a large sprayer was installed on the roof in order to spread the bacteria over Tokyo, which could of course have produced deadly pulmonary anthrax. This sprayer was brought into play in 1993, and in addition they drove their trucks around spraying anthrax on several occasions. The apparent lack of effect is attributed to their mistake in using a harmless vaccine strain. So extreme incompetence prevented a much worse outcome. The cult also tried to obtain the deadly Ebola virus, when their so-called humanitarian mission 'African Salvation Tour' of 40 people went to Zaire in 1992, but it is not known if they did collect any Ebola samples. Ian Reader's analysis of Aum has suggested that one reason that they were allowed to progress so far was because the Japanese authorities were nervous of being accused of persecuting religious groups as a result of their notorious pre-war record.

The killing of Georgi Markov in 1978 is the best known and most notorious example of state-sponsored use of a toxin weapon, and was described at the start of this book. The story was partly unravelled because of a similar, but unsuccessful, attack on another Bulgarian, Vladimir Kostov, at the Arc de Triomphe metro station in Paris in the same year. In this case the pellet* lodged under the skin of his neck, and insufficient toxin escaped rapidly enough to kill him. Some sources have suggested that, because Kostov survived, his body developed an immune response to the foreign protein and scientists were

---

*The pellet in Kostov's neck was one hundredth of a millimetre bigger, at 1.54 mm.

able to identify antibodies to ricin in his blood. None of the official reports carries this story.

It is thought that, after being filled with ricin, the holes in the pellets were sealed with a wax that would melt at body temperature to release the potent poison. Although the scientific aspects of this distressing story were identified within six months of the act, the people involved were never caught. After the breakdown of the Soviet Union, Oleg Kalugin, who had been in the KGB, suggested that the ricin had been made in Russia and given to the Bulgarians with the approval of Yuri Andropov, at the time the Head of the KGB. Following the fall of the Communist regime in Bulgaria, the Bulgarian authorities collaborated with the British to try to track down those responsible. Several senior officials of the Bulgarian army were investigated about the destruction of official documents. Francesco Guilino, a Danish citizen originally from Italy, has been suggested as the operative who used the poisoned umbrella. Guilino was questioned but released and no one has ever been prosecuted.

Whether this type of assassination was used on anyone else is controversial. The story that bears most credence concerns Boris Korczak who was allegedly attacked in the USA. According to Seth Carus, there are two versions of what happened to Korczak, but common to both is the suggestion that he was shot with a small metal ball of the type used against Markov and Kostov. At the time of the Markov killing, another Bulgarian died in London in mysterious circumstances. He had apparently fallen down the stairs. It is unlikely that we shall ever know the full details or the extent of the attacks carried out by the Bulgarian regime on its dissident citizens.

A small American tax protest group, calling itself the Minnesota Patriots Council, had plans in the early 1990s to use ricin to strike at local officials, but a tip-off meant that two of them (Douglas Baker and Charles Wheeler) were arrested before they could act. They were later convicted. Other such right wing groups have attempted to use ricin.

In September 2001, when America was still reeling from the attacks of 11 September, a new horror was unleashed—anthrax. This is a rare disease in the USA, particularly inhalational anthrax, and the only other occurrence of this version of anthrax in the previous century was in 1957 when four died in a factory that processed goat hair. The first case in September 2001 raised immediate alarm signals. Johanna Huden, who worked at the New York Post, developed cutaneous anthrax starting on 22 September. The source was never confirmed, but an envelope at her office dated 18 September was found to contain anthrax spores. The same pattern was repeated over the next two months, as more cases linked to the opening of contaminated mail were identified, leading to a total of 22 infected persons. The nature of the infection was divided equally between cutaneous and inhalational anthrax. However, all eleven cutaneously infected individuals survived, whereas five of the inhalational cases died. This bears out the concern that inhalational anthrax is very difficult to treat effectively.

The aerosolisation that led to the death of the postal workers was actually caused by some of the machinery that compressed the batches of envelopes, whereby air and anthrax spores were expelled from the envelopes. The anthrax attack also showed how widespread terror could be induced by a relatively small threat. The scale of the coverage and the public anxiety were totally out of proportion to the scale of the attack. Hoaxes and genuine scares exacerbated its importance and led to microbiology laboratories being inundated with requests to analyse samples, the overwhelming majority of which turned out not to contain anthrax. At the time of writing, the person or people responsible have not been identified. Indeed, there appears to be remarkably little about this story in the news now, and it has been suggested that the strains of anthrax bacteria used were of American origin.

Information on the cases described above is readily available, but it is not known whether this represents a true picture of the use of the terrorist biological weapons or whether this is a small fraction of a much larger and more subtle abuse of biology.

Several individuals have used toxins in straightforward criminal activity. The first documented case was in 1910, in St Petersburg. Patrick O'Brien de Lacy and Vladimir Pantchenko were convicted of the murder of Vassili Buturlin, who was de Lacy's brother-in-law. De Lacy wanted to ensure that his wife inherited as much as possible from her wealthy father, whereas Pantchenko was an unscrupulous doctor who was willing to kill for money. Although their first idea was to infect Buturlin with cholera, which was endemic in St Petersburg at that time, Pantchenko later decided to inject the unfortunate Buturlin with diphtheria toxin. He died seven days later.

Over the years others have used *S. typhi*, diphtheria, tuberculosis, and *Yersinia pestis* for financial gain or revenge. A particularly interesting case in the mid-1960s concerned a Japanese doctor who used *Shigella* and *S. typhi* to infect over 100 people, leading to four deaths. Mitsuru Suzuki infected colleagues because he felt that he was not being treated appropriately; he also infected members of his family, because he believed that they were not supporting him sufficiently financially. What is most amazing is that the Chiba University Hospital authorities had known about his actions for a year, but had kept them secret to avoid embarrassment to the university! He was never charged with murder.

More recent cases have occurred in America. In 1995, Dr Debora Green of Kansas bought some castor beans at a garden shop with the intention of poisoning her husband. He survived, but suffered several health problems. She clearly had psychiatric problems and never admitted that she had tried to kill him. However, antibodies to ricin were found in his blood and she was sentenced to 40 years. The following year in Texas, a laboratory technician called Diane Thompson was accused of poisoning people, in particular an ex-boyfriend whom she had pursued vindictively. She had used *Shigella*, which she had access to at the laboratory. Thomas Leahy did not appear to have money or revenge as a motivation for his activities, but was obsessed with poisons and the possibility of killing people. He had ricin in his possession and may have been looking to make other poisons. Although he did not apparently hurt anyone, he was sentenced to over 12 years in prison in Wisconsin for possession of ricin and intent to use it as a weapon.

The UK has also been subject to threatened attacks. In 1995, Michael Just, an Austrian with a microbiology degree obtained in Britain, tried to extort money from several British dairies by threatening to put *Yersinia* in their milk. He was caught as he tried to withdraw the money that had been paid by the dairies. There have been several other cases involving possession of ricin, or threats to use ricin or anthrax.

It is perhaps not surprising that most of the cases using biological agents as weapons have involved people with the relevant knowledge and potential access to such agents. There will always be those who exploit knowledge for their own selfish means, and biological weapons just give another opportunity for those with the expertise.

Biological weapons are not going to go away, whatever legislation is enacted, and however stringently some countries might try to enforce it. The agents viewed as being the most dangerous and plausible as weapons are infectious. Attempting to predict which of these agents is most likely to be used is very difficult. However, the viral disease smallpox appears to be at the top of every list. Its eradication in a WHO-led programme may have been a triumph for medical science, but as people are no longer vaccinated and thus highly susceptible to this very dangerous disease, it was seen by others as an opportunity for military use. Most importantly, the user's own people can be vaccinated, but the enemy will be susceptible. Smallpox is highly infectious and is relatively robust for a virus—it can survive on surfaces for an extended time, ready to infect others. The natural infection is fatal in about 50 per cent of cases, and the only real protection is by prior vaccination. This statistic could be made worse by manipulating the genome of the virus so that its surface proteins were changed in such a way that the vaccine did not raise a protective response.

Several bacterial pathogens are thought to pose considerable threats. Such bacteria could be made more dangerous by genetic manipulation so that they could evade control by vaccines or antibiotics, or by the addition of further toxin genes. Plague and anthrax each has the capacity to strike fear into a population. The ability of anthrax spores to lie dormant for extended periods of time is a bonus for the bioweapons' designer. However, only five people died in the 2001 US outbreak. Although each of these was a personal tragedy, this number is totally insignificant compared with the premature deaths caused by road traffic accidents, smoking, and natural diseases. It is important to keep the risk in proportion*—life is a dangerous occupation anyway.

The list of bacteria and viruses that might be considered as potential threats is a long one, and no one can be certain whether much of

---

*Harry Smith, who showed that anthrax was a toxin disease in the 1950s, considers that the risk from anthrax may be exaggerated, because humans are relatively resistant to anthrax.

this is hypothetical or whether some of the agents have been made, or genetically manipulated. The bacterial diseases tularaemia, brucellosis, Q fever, and typhus, the viral disease tick-borne encephalitis, and the haemorrhagic viral diseases, such as Ebola, have also been considered as potential weapons. Of course, the worst scenario is for the release of an agent for which no one has an effective remedy.

Non-infectious biological agents are not going to wipe out the world population or even a nation, but they can still cause widespread panic and disruption, not to mention death. These are really sophisticated chemical weapons. Botulinum toxin and ricin are the best-known options for weapons. Each toxin could be loaded on to food or the water supply, although the amounts needed would be very large. It is not known why the aerosolisation of botulinum toxin by Aum Shinrikyô failed to cause damage, although one possibility is that that strain was not particularly toxigenic.[8] The delivery strategy for these toxins would need to be well researched, although it is possible that secret work on this has already taken place.

Some bacteria and toxins could be used to incapacitate rather than to kill directly, so giving an enemy a short-term advantage. The staphylococcal enterotoxin B, a superantigen, has been suggested as a 'useful' weapon for this purpose. As a very common cause of food poisoning, it could also be hard to detect. The incapacitating dose is calculated as 0.4 ng/kg body weight,* whereas the LD_{50} is only 20 ng/kg body weight.† When inhaled this toxin causes fever accompanied by respiratory distress, nausea, and in some cases vomiting. Such a weapon could be useful in a battlefield situation, and the USA prepared it for such a weapon in the 1960s.

Non-bacterial toxins have also been considered for use as biological weapons. These include the potent neurotoxin saxitoxin, found in shellfish, which blocks sodium channels in cell membranes. Several marine toxins have similar properties. Fungal toxins, such as aflatoxin, have also been considered. Some of the snake venom toxins might be used as biological weapons, but are thought to be difficult to produce in large amounts.

*A nanogram (ng) is one-billionth of a gram.

†LD_{50} stands for the 50 per cent lethal dose, and indicates the dose that will kill 50 per cent of a targeted animal or human. This is considered the most accurate way to assess the toxicity of a substance, whether chemical or biological.

What defence is there against such weapons? Defensive programmes have remained in several countries, although there was a widely held view that such weapons would not be used because of basic human morality. History has taught us that basic human morality is a very precarious concept. However, governments became increasingly concerned from 1995 onwards, the year that the Aum Shinrikyô sect was found to be experimenting with biological weapons. In the same year, the Iraqi government admitted to a greater extent than before the scope and magnitude of its bioweapons' programme. Also, in that year, the US government began to take seriously the threat posed by the Russian weapons' programme, which employed about 50,000 people. The stockpiles of agents held were put in the range of many tons. These included smallpox, plague, anthrax, and various toxins. The dispersion of highly trained and intelligent scientists from the Russian programme to other countries remains a matter of concern. The revelations of 1995 led to a great expansion of various initiatives to try to deal with this threat, along with an immense injection of money. Research was greatly expanded and local public protection programmes were put into place, hopefully to deal with any situation that could be envisaged. Only time will tell if the threat was real and if the defences were appropriate.

The spectre of biological weapons sometimes whips up an anti-science response from those who wish that science had not led to an understanding of toxins and infectious agents. The corollary is to suggest stopping all such research, or at least preventing publication of the findings, lest they are taken up by rogue states or terrorist groups. After the scare of the US anthrax attacks in 2001, some suggested that publications on toxin research should be especially censored, but after due consideration it was decided that the advantages of making this research public far outweighed any real or perceived dangers from it.

The argument for wholesale censorship is untenable from several points of view. Natural infections over the centuries have killed many more than have been killed by biological weapons. Indeed, medical and scientific advances, based on a deeper understanding of viral and bacterial pathogens, have helped hold back the slaughter wrought by these pathogens—most significantly so far in the developed world. Those who oppose science and research because of its dangers, or potential dangers, are always vague about what 'natural' time they want to go back to. Was the Victorian era, with its appalling life expectancy and many common childhood illnesses, better? However far the anti-science lobby would like to take us back, we will still find dual-use technology. Science and technology have been with humankind since humans began to think. The branch plucked from a tree could be used to kill a neighbour, or could equally be used in hunting for food or building a shelter. Technology cannot be stopped or ignored. Openness and an informed public who can debate these serious issues sensibly will dispel the panic and misinformation that serve no one. Although we must not be complacent about the possible use of bioweapons, a rational view based on the facts is surely the way forward.

8

A MORE OPTIMISTIC OUTCOME
From poison to cure and the cell biologist's toolkit

For the pessimist, the last three chapters have painted an increasingly gloomy picture of potent and deadly molecules that possess within their molecular architecture an inert but uncanny knowledge of the workings of our bodies. These can either cause fatal disease naturally, or can be used by the madmen of the world as vicious and terrifying weapons. The optimist would have seen some hope as the nineteenth- and twentieth-century scientists gained an understanding of these molecules, and would be cheered by the observation that the use of bioweapons is being curtailed by most countries. However, there are highly positive aspects to these molecules. Not only have therapies been devised to protect against toxin action, but recently toxins have been used in novel and wholly beneficial ways.

The story of diphtheria showed how knowledge about a toxin could be used to make a highly effective vaccine. Over the years, a similar approach has been tried for other toxin-related diseases with varying success. Generally the more straightforward the disease process, the better the vaccine has been. Like diphtheria, tetanus was also tackled very early on in the history of microbiology. In each case the disease is clearly linked to a powerful toxin that alone causes all the signs and symptoms of the disease, so that its inactivation produced a vaccine with few side effects that is still highly effective and in use many decades later. This indicates that natural selection has not been able to select active toxin variants that can avoid being recognised and inactivated by the antibodies induced by the vaccine.

The perfect vaccine would induce long-term immunity against all possible variants of a pathogen after a single oral administration without inducing any side effects, would be cheap, and could be stored at room temperature. Such a vaccine would trigger immunity at mucosal surfaces* where most pathogens attack, and at the same time avoid the need for needles[1] or fridges that can both cause particular problems in the developing world. This is a very difficult goal to attain. Vaccines are therefore often a compromise, and assessing the value of a vaccine can be complicated.

Safety is an interesting concept. A vaccine, or indeed any medicine, that causes side effects to a small but significant number of immunised people would perhaps appear to be unacceptable. However, if the disease against which the vaccine protects is often fatal, or leads to long-term complications, it might be worth taking the risk, especially if you are likely to catch the disease. The first formulations for a diphtheria vaccine were quite dangerous, because they were a mixture of active toxin and anti-toxin antibody, but the disease itself was far more dangerous than the risk of the vaccine. The key issue is to assess the relative risk from the disease against the risk from the vaccine. A further difficulty arises in trying to persuade people to accept something that might affect them adversely, as opposed to taking chances with 'fate'. There may also be a reluctance to accept a vaccine that has to be given in multiple doses if there are temporary side effects that have to be experienced several times.

Although the ideal vaccine will induce long-term protection, a vaccine that protects for only a short time may be very useful in the face of an epidemic. The current cholera vaccine is a prime example of an imperfect vaccine that can still be effective in some situations. Similarly, if the pathogen has multiple types, a vaccine that can protect only against a limited number of types may still be useful if it protects against the most common types.

Vaccine cost is a problem. A highly effective vaccine that protects after one dose may potentially not generate as much money as a poorer vaccine that needs multiple doses. However, whatever our view of pharmaceutical companies, vaccines do cost a vast amount to develop and there are many stages to the process. Assuming that the infectious agent is known, bacterial components that might induce immunity have to be identified and then inactivated, and these have

---

*The mucosal surfaces comprise the respiratory, gastrointestinal, and urogenital tracts.

to be combined with other chemicals that will boost the immune system. There follows extensive laboratory testing, and then complicated, expensive, and detailed testing in the field. There will be many blind alleys—products that failed at some stage in the development process. The company involved has to make a profit, not only for the shareholders—who have risked their money—but also to finance the next potential vaccine.

It is interesting that any profit by a pharmaceutical company is often seen as intrinsically bad, as if the managers and scientists involved in the discovery and development of the product should not be rewarded, although similar considerations are not usually applied to doctors. Perhaps this is because they are seen to be directly delivering 'patient care', although it could be argued that the scientists within a pharmaceutical company play a substantial part in 'patient care', although from a more remote perspective. Perhaps the key moral element in this difficult issue is whether the vaccine companies make an excessive profit.

In 1974, the World Health Organization (WHO) launched the Expanded Vaccination Program to increase the numbers of children immunised worldwide. This was widely sponsored (Unicef, governments, banks, charities, etc.), with vaccine prices being tiered from poor to rich countries. This has been highly successful. It was estimated that, by 1990, 80 per cent of the world's children had been immunised against six target diseases,* compared with 5 per cent in 1974. However, attempts to add further vaccines to this list have met difficulties, partly as a result of the cost of new-generation vaccines.

Many vaccines have been made since the highly successful diphtheria vaccine. Although the tetanus vaccine is also very effective, tetanus still kills an alarmingly high number of people each year—up to half a million babies, and also several tens of thousands of unprotected mothers. A two-pronged approach is being adopted—an improvement in the uptake of tetanus protection and improved birth practices to prevent infection at this susceptible stage. Although the conventional tetanus vaccine is highly effective, it relies on booster doses being administered and this can lead to a high drop-out rate in some remote areas of the world. Novel approaches to combat this

---

*Diphtheria, tetanus, whooping cough, polio, measles, and tuberculosis.

have included the use of tetanus toxoid in beads to deliver the vaccine at timed intervals and a genetically modified live vaccine. In this latter vaccine, part of the toxin gene is stitched into a live but crippled strain of *Salmonella* that might induce better immunity from a single dose. The WHO has set a target of 2005 for the elimination of tetanus in the newborn.

Vaccines based on the chemical inactivation of a bacterium that produces toxins have also been effective. The vaccine against *Bordetella pertussis*, which causes whooping cough, comprises heat-killed bacteria, often administered in combination with tetanus and diphtheria. There have been various scares about this vaccine. A mild reaction to the vaccine was quite common, but there was also apprehension about very rare neurological complications. Large-scale surveys have not found a link between the vaccine and any permanent effect in children. Nevertheless these scares led, in the UK, to vaccine coverage dropping from 75 per cent to 25 per cent in the mid-1970s. As whooping cough is often fatal or can lead to serious complications in very young children, this is a case where the slight risk of the vaccine is worth taking. Even in older children whooping cough is an unpleasant disease.

Although about 80 per cent of people are immunised, whooping cough still affects around 30 million annually, with a death toll of around 1 per cent. This is a great improvement, but still unacceptable for a preventable disease. Since the 1990s vaccines have been available that contain just part of the bacterium and not the whole bacterium which is thought to lead to the side effects. These new vaccines contain the pertussis toxin in an inactivated form plus some of the adhesins—molecules that enable the bacteria to stick to our cells. These vaccines are thought to cause fewer side effects and are as effective as the conventional vaccine. It is not yet clear whether the immunity that they generate lasts as long as that generated by the conventional vaccine, and they cost more. Both issues are major concerns for a disease that is now mainly restricted to poorer countries.

Cholera was a bacterial disease desperately requiring a remedy during the nineteenth-century epidemics. It is not a disease that has gone away. In 1991, outbreaks in South America and Africa killed over 18,000 people. Some three years later, it swept through refugee camps in Rwanda and was on the increase in both eastern and western

Europe. As it is thought that only about 10 per cent of cases are ever reported, this remains a fearsome disease, with global estimates of 120,000 deaths annually—mainly children in the developing world.

Although Robert Koch had isolated the bacterium responsible for cholera, the understanding of cholera was complicated by Max von Pettenkofer's infamous and personal experiment of drinking a culture of the bacteria and remaining healthy. Many did not really believe that the disease could be entirely explained as an infectious one. A Spanish attempt at a vaccine, although apparently beneficial, had not helped the cause. This was because the scientist involved, Jaime Ferran y Clua, refused to give all the details to a Scientific Commission from the Institut Pasteur.

The development of a cholera vaccine was then taken up by Waldemar Haffkine, a Russian émigré, who had various personal and scientific obstacles to overcome before he was successful. Early in his career, Haffkine was befriended by Ilya Metchnikov (whom we met earlier), although later on Metchnikov was less than helpful in promoting cholera immunisation, as he also believed von Pettenkofer's ideas. One of the problems with cholera was that it caused disease only in humans. Haffkine first grew the bacteria in the laboratory with increasing amounts of rabbit serum to toughen them up (presumably by the acquisition of mutations), and then infected animals with the resulting bacterium. It had become more virulent after growing in the animal. The final product represented the vaccine, which was administered by injection under the skin. It is interesting that the principles behind this vaccine were completely different to the diphtheria, tetanus, and rabies vaccines where the need was to make the pathogen weaker. The other diseases operated in the whole body so that injection of the pathogen under the skin would cause the disease; cholera happens only in the gut of an infected animal, so the bacteria were not able to survive long enough to raise an immune response under the skin unless they had been made stronger. As Europe was not willing to support the concept of a cholera vaccine, Haffkine travelled to India, where in 1894 he was able to show that it worked.

It has not proved easy to make a very effective vaccine against cholera. The conventional vaccine that is injected has been around for over 40 years. It is a chemically killed vaccine, but is known to be less than 50 per cent effective, and the immunity induced usually lasts less than six months. Moreover this vaccine does not prevent spread, and latterly its use has not been recommended by the WHO. It is important to improve the vaccine, because cholera is still a major killer.* Yet even the understanding of the action of the toxin and the availability of molecular tools do not make the job easy. Mutant strains of *Vibrio cholerae* that do not have the toxin gene still cause diarrhoea, albeit a mild variety of the disease. This is partly because *V. cholerae* makes at least three other toxic proteins, including an extremely large pore-forming toxin.

New cholera vaccines that can be given orally have recently become available. These appear to be promising and can be used in an HIV-positive population. One of these is a killed vaccine to which the cholera toxin-binding subunit (the B domain) has been added, and trials with this have shown 85 per cent efficacy after six months and even 50 per cent after three years. As a result of the close similarity between cholera toxin and the LT enterotoxin of *Escherichia coli*, this vaccine also gives some protection against this other diarrhoea-causing bacterium. The other vaccine is a live one that has been weakened by genetic manipulation; it is also effective and safe.

A novel and exciting option now being investigated is to transfer the DNA for part of the cholera toxin B domain into plants that can easily be grown in these countries. Eating the genetically modified (GM) plants would provide protection. This would be of great benefit for the countries where cholera is a problem, because even relatively cheap vaccines can be a drain on limited health resources. The safety of any vaccine in HIV-positive individuals is also important. The WHO has analysed the relative effectiveness and the cold economic advantage of vaccines depending on the likely incidence of cholera during epidemics. Prevention remains the best way forward, but, in some cases, the simplest aspects of intervention, such as disinfecting a contaminated supply, can have a major outcome. As is well known, the simplest treatment for this disease is rehydration with a solution of clean water containing a balance of salts and glucose. Antibiotics should be used only for severe cases, and their overuse has already led to antibiotic resistance.

Bacteria that deliver their effector proteins by the specialised type 3

---

*The world is currently in the seventh cholera pandemic, which began in 1961.

or 4 secretion systems directly into our cells are of course major killers and vaccinologists have worked hard to combat these serious diseases. Vaccines against these pathogens have generally been made, not against the toxic proteins, but against either the whole pathogen or carbohydrate components from them. This is probably because knowledge about this group of toxins is relatively new—probably too new to be incorporated into vaccine strategies. In addition, as the effector toxins are directly injected into our cells, they are never outside the cell and may be protected from attack by the immune system.

Efforts to combat plague began in the 1890s and were started by Haffkine, who had also pioneered cholera vaccines, and who was still in India when Bombay was hit by an outbreak of bubonic plague. Haffkine developed a heat-killed vaccine containing both the bacterium and the soup in which they had been growing. His vaccine was successfully tested in a prison. Haffkine then suffered a number of injustices at the hands of the (British) Indian government who were more interested in using sanitation to combat plague than a new-fangled vaccine. When, however, he had an opportunity to prove that his vaccination approach was successful, the British decided to replace him with one of their own people. Their chance came when a contaminated vial of vaccine caused the death of 17 people. Although Haffkine was not to blame, it was suggested that he return to England. His reputation was rescued by a letter to *The Times* signed by Ronald Ross* and other illustrious scientists. Haffkine then returned to India where he was still treated rather dismissively by the authorities there. The USA produced a modified heat-killed vaccine based on Haffkine's work in the 1940s, although there has been substantially more interest in vaccines against plague in the last few years.

Typhoid remains a major worldwide killer. An effective vaccine based on killed *Salmonella typhi* bacteria has been available for some time, but it requires multiple doses and protection is short-lived. It was therefore recommended only for travellers to regions where typhoid could be contracted. Newer vaccines include the live attenuated Ty-21 strain which has been subjected to long-term tests for safety and effectiveness. Ty-21 appears to be as effective as the

*Ross was famous for his work on malaria.

conventional vaccine, but without the side effects. Other genetically manipulated vaccines that are attenuated are under development, as well as those based on polysaccharides (complex sugar molecules) extracted from the bacteria.*

Shigellosis is a killer worldwide on the same scale as typhoid, which is causing alarm because of the appearance of antibiotic resistance. Both polysaccharide and live attenuated vaccines are being developed to combat it in laboratories in the USA, France, and Sweden, and several of these are currently being tested for their effectiveness.

There is also a worldwide effort to make vaccines against many other important and deadly diseases, including those caused by bacteria such as *Mycobacterium tuberculosis* and the streptococci, as well as dangerous viral diseases such as measles, polio, respiratory syncytial virus, rotavirus, the hepatitis viruses, and of course HIV/AIDS. In addition, the parasitic diseases, malaria and schistosomiasis,† annually kill at least 1.5 million and 200,000 respectively, and improved control measures are desperately needed.

The current increased concern about bioweapons has accelerated the drive to make vaccines to protect against terrorist attack. The original Pastorian live vaccine against anthrax was slightly modified in the early twentieth century, but has largely been regarded as effective. However, as there is still some risk of infection from this vaccine, it has not generally been used in humans. Vaccines based on the protective antigen, the binding domain of the toxin complex, have become available for human immunisation. These are very effective against the cutaneous form of anthrax, although annual boosters are required and there are mild side effects. This immunisation is recommended for people working in trades where they might come into contact with anthrax, such as those working with animal skins. Obviously more effective vaccines are required. From the time of the Sverdlovski incident in 1979, there has been urgency in improving anthrax vaccines, with extra impetus being provided by the US postal system attacks following 9/11. Several approaches are being used: recombinant protective antigen (PA) modified to stimulate the immune response better, recombinant PA expressed within a host

---

*Similar types of vaccine are used to combat meningitis.
†Schistosomiasis also predisposes towards bladder cancer.

bacterium, and mutated *Bacillus anthracis* which is not pathogenic, but is still able to stimulate a good immune response.

Vaccines against botulinum and the *Clostridium perfringens* toxin have also been prepared and the Russians are known to have produced a live attenuated vaccine against tularaemia in the 1950s. A very substantial effort is currently being made to generate improved vaccines against these and a number of other potential bioweapon threats.

The constant interplay between synthetic vaccines and the bacteria that seek to colonise us depends on a number of factors. A wary public, now seemingly sceptical of any scientific advance, demands safer and safer vaccines. Although it is reasonable to demand that vaccines should be as safe as possible, it is a dangerous situation when people would rather have no vaccine than one that might have side effects, without assessing the relative risk. This view is all too often fuelled by a strident press, desperate for headlines and not too concerned by application of the facts. If the bacterium is able easily to change the part that is targeted by the vaccine without losing its ability to cause damage, the vaccine may not be very effective for long, and new vaccine formulations will be needed. On top of this, as we are still being bombarded by new diseases, or diseases that we now recognise as being infectious, the quest for new vaccines is very unlikely to abate.

Toxin research has always been at the forefront of biology. In the nineteenth century, work on toxins had helped to define and establish microbiology, immunology, and epidemiology. Later the failure to find bacteria in what were clearly infectious diseases would lead to the discovery of viruses. However, the potency and precision of toxin action have more recently been applied in a far wider arena of biology—that of cancer.

Even before bacteria were discovered, it was known that acute infections could sometimes cause tumour regression. In 1868, before the link between bacteria and disease was recognised, the German surgeon Wilhelm Busch infected some cancer patients by exposing them to the skin disease erysipelas. The improvement was slight and only temporary. Twenty-three years later, by which time it was known that *Streptococcus pyogenes* caused erysipelas, the New York surgeon William Coley took the next step. Desperate to find a better treatment than surgery after the unpleasant and rapid death from cancer of an 18-year-old girl,[2] he searched the available literature and found several examples of infection being used to treat cancer. One patient whom he tracked down had lived healthily for seven years after treatment. Over a long career Dr Coley carried out a thorough investigation of the effectiveness of such treatments using different bacteria. The addition of *Serratia marcescens*, then known as *Bacillus prodigiosus*, to the formulation enhanced the effect and the two bacteria, both heat killed, came to be known as Coley's toxin. Some tumour types responded better than others. On Coley's death, his daughter Helen Coley Nauts, who was not medically trained, sorted and collated his records to confirm that the treatment could be effective. In 1953 she founded the Cancer Research Institute to investigate why and how Coley's toxins worked. It is now realised that the therapy stimulates the immune system to attack the tumour and there is a resurgence of interest on this theme. In 2004, the Cancer Research Institute, along with the Ludwig Institute for Cancer Research, signed a contract with Cobra Manufacturing plc for the commercial production of Coley's toxin for clinical trials.

A more direct and targeted attack on cancer came with the use of molecular biology to link the potent cell-killing capability of a toxin with another molecule that could recognise and bind specifically to a cancer cell.[3] The resulting chimaeric molecule should be restricted to attacking only those cells that display the cancer-specific marker. Conventional cancer treatments, although designed to attack cancerous cells differentially, are never 100 per cent efficient. This happens in both ways, because these treatments attack some non-cancerous cells, but cannot be relied on to deal with all cancer cells. Damage to healthy cells is a nuisance that causes side effects. However, more significantly, if some cancerous cells remain, the cancer will reappear. These problems can arise with any form of therapy, including surgery, radiotherapy, and chemotherapy.

Toxin chimaeras have been generated using several different toxic activities, coupled to a wider variety of targeting molecules. The resulting chimaeras are frequently referred to as immunotoxins because the targeting part of the molecule is usually either an antibody that is specific for a cancer marker or a molecule involved in immune function. The toxins that have been mainly used for immunotoxins are diphtheria toxin, exotoxin A (ETA) from *Pseudomonas aeruginosa*, and the plant toxin ricin. In each case a tremendous amount of information is known about these toxins at the molecular level and the structure of each one has been worked out at the atomic level.

The hope that this approach could rapidly provide a smart approach to cancer treatment was soon dashed by setbacks. Both parts of the immunotoxins were recognised as foreign by the immune system, and any immunotoxins made with the diphtheria toxin were a particular problem as most people have some immunity to diphtheria. The generation of immunity to the antibody part of the immunotoxin could be avoided by using just part of the molecule, but the immunity to diphtheria has been harder to overcome.

The immunotoxins turned out to be less specific than had initially been hoped. In the next generation of such molecules, the toxin domain was linked to one of a number of small molecules such as cytokines. Cell surface receptors for many of these small molecules are often present in high numbers on cancerous cells. Interleukin-2 (IL-2), which is involved in many aspects of immune function, was coupled to either the diphtheria toxin or the *Pseudomonas aeruginosa* ETA domains. Unfortunately, these chimaeric proteins still attacked healthy cells, probably because such a small amount of the chimaeric protein actually reached its target, the rest being mopped up by other cells.

Despite these setbacks, some of the chimaeras have now reached advanced levels of development and are being tested in clinical trials. It is thought that blood cancers are particularly susceptible to immunotoxin attack. As immunotoxins do not penetrate far into tissues, they are likely to be less effective against solid tumours than against individual cancer cells found in blood cancers. It was perhaps naïve to expect that these chimaeric molecules would instantly provide powerful, effective, and highly selective therapy with few side effects. Much has been learned in the 15 years or so of research on these molecules and there remains great hope that this approach will live up to its earlier promise.

In a different approach, the lethal factor of anthrax toxin is being investigated for anti-tumour activity. This protein attacks a key signalling protein that is often mutated in human cancers. The injection of lethal factor into tumours in experimental animals greatly reduces their growth. The anthrax protein that binds to the cell surface, protective antigen, also has to be injected for this to work. A clever trick is now being devised to target this lethal combination to cancer cells. The protective antigen is normally cut with a protease present in all cells, and this is necessary for it to be able to bind the lethal factor and ferry it into the target cell. As some tumour cells make proteases that are specific for that tumour, scientists have mutated the protective antigen so that it is now recognised only by the protease found in the cancer. This mutated protein will be able to bind to lethal factor only if it has been activated by binding to a cancer cell with the specific protease.

It is likely that more of these elegant therapies using toxin molecules will be devised in the future. Over the last 10 years or so, people have been thinking of ways of attacking cancer cells much more directly, using the even more specific properties that make them cancer cells, that is the mutations that led to the cancer. That type of therapy would be very specific indeed, and it is possible that within the next 50 years all current treatments will be superseded by much smarter remedies that target only cells with cancer mutations that are never found in normal cells. Until that stage is reached, therapies exploiting the powerful properties of toxins are likely to have a significant part to play in the fight against cancer.

The use of the most deadly bacterial toxin to remove the natural wrinkles of ageing is well known. However, botulinum toxin was first used to treat a whole variety of real medical conditions. As we saw in Chapter 5, Justinus Kerner had suggested, in the 1820s, that botulinum might have a therapeutic role in blocking nerve function. Starting from the 1980s, botulinum toxin has been applied therapeutically for conditions caused by excessive muscle contraction which is triggered by overactive nerve function. The botulinum serotype A toxin has been used for this work, sold under the name BOTOX®, or Dysport® in Europe. Local injection of very small quantities of toxin is directed just at the muscles causing the problem and can be used selectively to weaken these muscles. The first condition to be tackled was strabismus, or squint, where excessive tension in one of the eye muscles prevents proper focusing. A whole raft of other distressing conditions resulting from excessive muscle contraction has since been treated, such as blepharospasm (where the eyelids are permanently closed, rendering the patient functionally blind) and torticollis (where neck muscles are permanently contracted, causing twisting of the neck). The toxin is also effective in other situations linked to over-active nerve activity, such as excessive sweating. The list of potential uses for the toxin continues to grow, and it may even be effective against migraine.

There are numerous advantages to the use of botulinum toxin, and apparently few side effects.[4] The toxin dose can be regulated and the effects last for months. It is administered without the use of an anaes-thetic. Even where the response is slightly too great, the dispropor-tionate effect will eventually wear off. Some people do not respond to the treatment and it is assumed that they have anti-toxin antibodies. In addition some people raise an immune response to the toxin treat-ment, especially if the site of injection is near an immunological hotspot that can pick up the toxin as foreign. However, as there are seven different types (serotypes) of botulinum toxin, it is possible to use one of the others in this circumstance and several other serotypes are commercially available. Each serotype appears to display slightly different properties. In particular two of the serotypes survive in the body for a shorter time than serotype A because they are more easily degraded inside cells, and these serotypes are likely to be very useful for situations where a short-term inhibition of nerve activity is required.

An exciting development is the deliberate manipulation of the botulinum toxin so that it will specifically target other types of nerve cells. Researchers at Porton Down have modified the cell-binding part so that the resulting toxin inhibits nerves involved in the trans-mission of pain. This has tremendous potential for the many people who suffer from chronic pain.

Toxins have proved to be amazingly useful reagents for dissecting how the cell works. In many ways this aspect of toxin use overlaps with the applied uses discussed in the previous section, because basic information about cellular function can become translated into prac-tical uses. The great precision with which toxins interact with our cells has led to them being categorised as the cell biologists' toolkit.[5] This particularly applies to the toxins that act intracellularly, which have a predilection for choosing targets that are essential for cellular function.

Usefulness in cell biology is not just confined to these intracellularly acting toxins. Some of the pore-forming toxins are used to generate pores for delivery of active molecules into cells. The pore-forming toxin streptolysin O is often used in the laboratory to allow large molecules into the cell, to see how they perturb cell function. The procedure is to add the streptolysin O to the cells together with the substance to be forced into the cell, followed by placing the cells in a toxin-free solution so that the cell can repair the hole. Some of these types of toxins might be used therapeutically for drug delivery. Toxins that attack the cell membrane enzymatically have also been useful because these can very precisely remove components of the membrane, and the researcher can then assess how this has affected cellular function.

However, the greatest versatility of toxin use lies with the intracellularly acting toxins. The three separate functions that are often coded by three separate domains have each been exploited to probe cellular activity. Toxins first bind to the surface of cells by attaching very specifically to a cell surface molecule, usually either a protein or a lipid molecule. For example, cholera toxin binds exclusively to lipid GM1, and has been used to study the location of this molecule in the cell membrane.[6]

Toxin entry into cells is complex. After all, the cell is not designed to take up poisonous molecules, and it appears that toxins have commandeered mechanisms used by the cell for the export of molecules. As different toxins enter the cell by subtly different mechanisms, studying how toxins get into cells has provided a wealth of information about how cells export proteins and also about how they move proteins around the cell. Toxins, such as tetanus and botulinum, that attack only nerve cells can be used to look at transport specifically within these types of cells.

When a protein is made that is destined for outside the cell, this protein is trafficked (moved) through a complex system of membranes inside the cell. Many toxins move backwards down this pathway, so the process of toxin transport is often called retrograde transport. Depending on the toxin, it will leave this pathway at a particular point, when the pre-programming built into its molecular make-up determines that it has reached the correct place for movement across the membrane into the body of the cell, to do its malicious work. We have already seen how the composition of the diphtheria toxin mol-

ecule enabled it to sense the pH (acidity) at which it should trans-
locate across the membrane into the cell. Diphtheria toxin typifies
toxins that enter by the 'short route'. These toxins leave the retrograde pathway quite early on. Other toxins travel deeper into the cell
before moving across the membrane, using a 'long route' to gain
entry. As well as telling scientists something about how these toxins
work, this type of toxin is also being exploited to drive proteins of
interest deep into the cell.

One use for this knowledge is to manipulate the immune system
so that it can fight pathogens and tumours more effectively. When
immune cells take up parts of the invading foreigner they move these
proteins down a retrograde trafficking pathway and, depending where
the part of the invader exits, they can determine what type of immune
response is produced, and thus how effective it will be against a given
pathogen. So it is entirely possible that in the future vaccines can be
engineered to produce an immune reaction that is tailored for the
best attack on the pathogen or tumour.

As already mentioned, it is not easy for proteins to cross membranes, and specialised mechanisms have to be employed by toxins to
enable them to do this. Work with our old friend diphtheria toxin
showed just how much a protein has to change to be able to squeeze
its way across a membrane. When little pieces of protein were stitched
on to the end of diphtheria toxin, the chimaeric molecule stuck
halfway across the membrane if the stitched-on piece was unable to
open up. This work led to the idea that a toxin must unfold and cross
the membrane almost as a thread—like a ball of wool being passed
through a small hole by being unwound on one side and rolled up
again on the other. This knowledge is being exploited to use this type
of toxin to deliver molecules to the cell. That rather odd toxin, the
adenylyl cyclase toxin from *Bordetella* species, which looks like a catalytic domain that is hitched up to a pore-forming toxin,* appears to
jump straight into cells and is one of a number of toxins being looked
at as a possible vehicle for delivery.

Analysis of the type 3 delivered toxins that are injected into cells
has also provided information about trafficking. Many of the bacteria
that produce these toxins choose to live inside cells, at least for part of

*See Chapter 5 for more detail.

their lifecycle. However, if bacteria are engulfed by cells programmed to scavenge for foreign invaders, the surrounded bacteria are held in a membrane vesicle which is trafficked down a pathway of increasing acidity, and enzymes are placed into this vesicle to degrade the bacteria. Clearly this would not suit a bacterium that hankers after an intracellular lifestyle. Such bacteria manipulate the cell so that the vesicle falls off this degradative pathway—again, studying these bacteria tells us not only about the bacteria, but also about the cellular processes that they usurp.

The catalytic activities of intracellularly acting toxins have provided an abundance of information about the cell. The use of chemical inhibitors in cell biology is problematic because they may affect several components in the cell, some of which the researcher may not even be aware of, and interpretation of the result is therefore prone to error. Toxins, on the other hand, act as precise probes for cell function. Several toxins have been instrumental in identifying important signalling molecules. Study of the action of the cholera and pertussis toxins that chemically modify the G-proteins G_s and G_i, respectively (see Chapter 5), led to the identification of these important signalling molecules. Pertussis toxin in particular is routinely used to probe whether G_i is involved in a particular process. Similarly, the *Clostridium botulinum* toxin C3, which inactivates the Rho protein, led to the discovery of the Rho proteins—incredibly important proteins that now appear to be involved in almost every aspect of cellular function. C3, and indeed many of the other toxins that act on Rho, are routinely used to investigate whether Rho is involved in a particular process. As some of the toxins that operate on Rho have differential effects on the family of different Rho proteins, a further layer of subtlety can be employed. Although not an intracellular toxin, the small stable toxin (STa) from *E. coli*, which binds to the guanylin receptor on our gut cells, should also be mentioned again. It was research on this toxin that led to the discovery of this receptor, which is important in the regulation of water balance. The tetanus and botulinum neurotoxins act by degrading proteins exclusively involved in the docking of membrane vesicles on to the cell membrane, and so have proved useful in understanding that process. As other toxins are identified and understood at the molecular level, the cell biologists' toolkit will continue to grow in both size and usefulness.

Right from the start we have seen that understanding toxins has been closely linked not only to devising novel therapies to combat toxin-based disease, but also to probing how our cells work. Thus the nineteenth-century studies of toxins not only led directly to vaccines to counter these diseases, but also identified the existence of immunity. In the twentieth century, toxins enabled scientists to begin to glimpse into cellular function, and by the end of that century were being used as precise and unique tools for those interested in cell biology and the mechanisms of cancer. In addition, practical ways to use toxins for entirely beneficial purposes are increasingly being found.

The non-vaccine applications for toxins are relatively new, and novel approaches are still being devised. However, we are not only fighting back against the cunning toxins, but we are now taking advantage of their cleverness for other beneficial purposes as well.

9

WHERE IS TOXINOLOGY[1] GOING NOW?

Is there anything new out there?

It has been pointed out that humans have more or less dealt with the larger animals on the planet, so the lions and bears and wolves are no longer much of a threat. That leaves the most dangerous animal of all, man, and the microbes to cope with. Whether humans can survive without using their own clever technology to kill their own race is impossible to tell. Of the microbes, many would see the viruses as the greatest threat. Horrifying as the AIDS epidemic is becoming, it is unlikely to lead to worldwide extinction unless its mode of transmission changes. Viruses such as Ebola, flu, and smallpox are potentially more worrying. We should also not forget that it was a bacterial disease, the plague, that almost did annihilate the race during its first two terrible pandemics. We should not be complacent.

What is likely to happen next in the toxin story? As with other branches of science, research on toxins is likely to proceed at a steady pace with sudden leaps into new and unexpected areas. Predictions are always dangerous, because many may well turn out to be wrong, but it is intriguing to look at some of the exciting newer discoveries and breaking news.

The battle to generate new vaccines and to make these as effective as possible will continue for the foreseeable future. But novel methods of combating bacteria and their toxins will also be found. Recent research on anthrax, driven by the terrorist threat, shows several ways that this might develop. It is now known how anthrax controls its toxin genes and this opens up the possibility for designing a therapy to act to switch these off rapidly at the early stages of an anthrax infection or attack. Other approaches manipulate the protective antigen part of the toxin so that it cannot facilitate uptake of the toxic factors, the edema and lethal factors. In one of these schemes, a soluble version of the anthrax toxin receptor has been constructed that cannot insert into the cell membrane. When large amounts of this modified receptor are injected, they will bind to and mop up any protective antigen produced by the infection and prevent it from binding to functional receptors. Using a similar strategy, scientists have altered the protective antigen so that it cannot bind to the enzymatically active toxin proteins. Injection of large amounts of this crippled protective antigen will out-compete any protective antigen produced by the invading anthrax bacteria for binding to the cell surface receptors, and thereby prevent uptake of the toxic factors. Each strategy blocks the anthrax intoxication right at the start.

It is increasingly being realised that the pore-forming toxins display an unexpected subtlety in their interaction with their target cell. Listeriolysin, a toxin from the *Listeria* bacterium known to be in many cheeses, does not kill cells. Instead it affects a different membrane, an intracellular membrane. Listeriolysin makes a hole in the membrane of the vesicle that *Listeria* is in, inside the cell, to enable the bacterium to escape. How does it attack one membrane and not another? The *Helicobacter pylori* VacA toxin also appears to target internal membranes. We may find in the next few years that the pore-forming toxins have other effects in the cells aside from pore formation.

Toxins delivered by types 3 and 4 secretion are a relatively recent concept, and there are bound to be many more of these waiting to be discovered. A particularly remarkable feature of these toxic proteins is that some are not enzymes and instead mimic normal cellular signalling proteins that work by transiently binding to other signalling molecules. Perhaps these bacterial toxic proteins are closer relatives to signalling proteins in our cells, and as such they are likely to be useful additions to the cell biologist's toolkit. Bacteria are known to export proteins by yet other strategies and we may find novel toxins there too.

The directly injected toxins solved the conundrum of those very dangerous pathogens that did not appear to make toxins and showed us that *Yersinia*, *Salmonella*, and many other bacteria were indeed toxin factories. What about the remaining disease-causing bacteria, the intracellular bacteria such as *Chlamydia* species and *Mycobacterium tuberculosis*? *Chlamydia trachomatis* is the most common bacterial cause of sexually transmitted infection, and can lead to infertility in women, often without any other obvious symptoms. These bacteria also cause a very unpleasant eye infection, trachoma, most notably in Africa. *Chlamydia* bacteria are very small with an unusual lifestyle, and they used to be mistaken for viruses. Like viruses they have a metabolically active (using food and growing) form that is not infectious and found only inside cells, as well as an infectious form that is not metabolically active. Recent analysis of the *Chlamydia* genome sequence has identified possible toxin candidates in these bacteria that are similar to known toxins, but it is too early to say whether these are important in disease. The *Rickettsia* genus of bacteria also has an intracellular lifestyle and is best known as the cause of the disease, typhus, which killed millions in the last century. There is some evidence that these bacteria also release enzymes to attack the host, but even less is known about these.

Mycobacterium tuberculosis kills around two million people annually and is a well-adapted and heavy-duty pathogen. It has very recently been shown that the tuberculosis (TB) bacteria produce two molecules that enable it to escape destruction inside the macrophage. One of these is a glycolipid (made of lipid and complex sugars), whereas the other is a protein that acts as a phosphatase. These two proteins prevent the fusion of the vesicle containing the TB bacteria with a vesicle armed with acid and enzymes ready to attack it. It thus seems likely that any successful pathogen will manufacture its own powerful weapons to interfere with and regulate host cell function, in particular to modulate the immune system.

Toxins that have been known about for years can still come up with surprises. The *S. aureus* toxin identified by Philip Panton and Francis Valentine in 1932 did not attract much attention then because it was found in less than five per cent of these bacteria. The toxin attacks immune cells and is named the Panton-Valentine leukocidin (PVL). It is a pore-forming toxin comprising two different proteins and its genes are encoded on a bacterial phage. Most importantly, PVL has now been found in a very high percentage of community-acquired* cases of methicillin-resistant *S. aureus* (MRSA), causing skin infections and serious disease, such as pneumonia in healthy young people.

*As opposed to hospital-acquired.

New toxins that display novel ways to attack the cell are still being discovered, even from such well-known bacteria as *E. coli*. Given that there are thousands of different bacteria in the environment about which we know very little, there are likely to be many more toxins with novel activities yet to be identified.

The atomic structure of many toxins is now known and many more will be worked out in the near future, thus enabling a more detailed understanding of how toxins interact with their molecular targets.

The accelerating advances in molecular biology have enabled scientists to obtain the sequence of DNA molecules very rapidly, leading to the ultimate and intrinsically fascinating sequencing project to obtain the entire coding sequence for a human being—the human genome project. This massive undertaking published the 'gold standard' sequence in 2004, showing that each of us has around 25,000 genes—astonishingly only about tenfold more than most bacteria. This amazing achievement provides a database of tremendous potential for the analysis of human cell function in health and disease, although by itself the sequence will not enable us to understand everything about being human. The complete sequence of many viruses has been known for many years, including that of HIV. Yet this knowledge in itself does not enable us to understand, let alone combat, these relatively simple organisms. Although the complete gene sequence of an organism is important, it is only the first step to understanding how the individual proteins made by the genes work and interact with each other. It is still necessary to carry out experiments in the laboratory in this post-genomic era.*

Meanwhile microbiologists have been quietly getting on with the same process for their favourite bacteria. The complete DNA sequence is now known for over 225 bacterial species. The sequences obtained are analysed using computers that search for the trademark signatures that indicate the start and end points of genes. The next stage is to ask the computer to interrogate known sequences and this gives a list of probable genes that are similar to those previously found in other organisms. Then, computers can look for smaller DNA signatures that indicate a possible function, such as similarity to a protease. This will suggest possible gene functions for other genes, some of

---

*This rather inaccurate phrase is commonly used to describe present-day biology.

which are more likely to be correct than others. What is left is a collection of genes of unknown function. These genes are possibly the most interesting, because they are likely to be those that determine why a particular bacterium has its unique characteristics, but they are the hardest to figure out. They may have novel functions and it will take 'classical' molecular biology, and time, to solve what they do. Ultimately, understanding of the complete genome sequences of bacteria will lead not only to the identification of new toxins, but probably also to new types of toxins.

Although we now believe that most bacteria make toxins, it has always been assumed that viruses killed us by taking over our cells for the assembly process of making new copies of themselves. Viruses generally shut down host functions so that the infected cell becomes dedicated to viral production. However, some viruses may have toxin-like activities. Rotavirus, a cause of childhood diarrhoea, has been reported to have a toxic activity that alters calcium signalling in cells, but little is known about this activity.

Bacteria gain various advantages by producing toxins: getting hold of nutrients, inactivation of immune functions, and, particularly for bacteria that live inside cells, manipulation of the cell's ability to engulf and destroy bacteria. To do this, the bacterial toxins display a highly sophisticated understanding of the way the cell works. How did bacteria obtain this knowledge? This is not known, but it is possible to speculate. The genes for many toxins are on mobile genetic elements. These are pieces of DNA that can move around easily, such as plasmids and bacteriophages (bacterial viruses), which suggests that these genes have been acquired relatively recently by the bacteria. Pathogenicity islands are also thought to originate in a similar way and often show remnants of phage ancestry. One possibility is that these genes originate from our cells, or more likely from our distant ancestors, or distant ancestral neighbours, such as forms of life that are only slightly more complex than bacteria. In other words, bacteria may have picked up genes that normally function in our cells and then subtly adapted them. As more bacterial and other genomes are sequenced we may find further clues about the origin of toxin genes.

Several toxin genes are found on bacteriophages that are still functional and thus may be able to transfer to new species where the DNA can be fixed into the genome of a new host bacterium. Bacteriophages can also perform another function. We have seen that toxins use several different ways to get into our cells, and this diversity reflects the great difficulty proteins have in crossing a cell membrane. But before they reach our cells, toxins have to escape from the bacterium. An equally large number of ways of crossing the bacterial membrane have been adopted by toxins. Some toxins use the general secretion pathway, found in bacteria and also our cells, whereas others require their own dedicated proteins to enable them to cross the bacterial membrane. Some toxins do not appear to do either. Recently it has been found that some toxins on bacteriophages split open the bacterial cell to escape. In this situation not all the bacteria die, otherwise the species would become extinct. However, a proportion of the bacterial cells die to release their toxin, so that their sister bacteria can prosper in the improved environment generated by the destructive power of the released toxin.

Why are any bacteria pathogens? If a bacterium is so virulent that it kills its host it will lose its food supply. It could be argued that a better lifestyle would be for a bacterium to cooperate with its host, or at the very least not harm it. That way the bacterium will not get itself harmed. Many bacteria adopt this strategy. Indeed, evidence in support of this argument is that new pathogens often become less virulent with time, because pathogenicity is not a wholly satisfactory lifestyle. However, bacteria deal in short time spans. Any bacterium that accidentally acquires a toxin gene has a short-term advantage over its relatives, because this growth advantage will lead to rapid growth and expansion of its own numbers at the expense of other bacteria. This strategy may not be in the long-term interest of the bacterium. In a parallel way, a cancer cell also has a short-term advantage but no long-term survival strategy, because unchecked it will kill the rest of the organism and thus its source of food.

A strange parallel can be discerned between the last 25 years of the nineteenth and twentieth centuries. The nineteenth-century theories about infectious disease, the most deadly of its age, were at best very hazy until that leap in understanding that began with Robert Koch in 1876. That new knowledge led, by the end of that century, to the definition of many causes of infection and a fundamental appreciation of what infection was. Rational therapy to combat those infections was devised almost immediately in the form of specific vaccines and chemicals. The first of these therapies was crude and produced what would now be regarded as unacceptable side effects, although at the time such risks were worth taking. However, as knowledge increased, the therapies improved, in terms of both their effectiveness and the induced side reactions. Later the discovery of antibiotics provided a very targeted therapy based on the differences between the properties of the bacteria and those of our own cells.

In a similar way there were several seemingly contradictory theories of cancer, the most feared disease of its time,* in the early 1970s. The last 25 years of the twentieth century led to a detailed and fundamental grasp of what cancer was, and thus the beginnings of more rational therapies to combat it. Many of these treatments still produce unpleasant side effects. The hope is that the present relatively crude strategies for attacking cancer can be replaced by therapies that are more specific in targeting only the aberrant cancer cell, and that will leave healthy cells untouched. Approaches along these lines are already being planned, such as therapies that kill only cells that have lost the ability to commit cell suicide.

In each century, the big advance was made when several, apparently unconnected pieces of information were brought together. The belief in the germ theory of life set the scene for the dawn of microbiology, whereas culture and staining techniques, improvements in microscope technology, and the serendipitous choice of which bacterial disease to investigate (anthrax) jointly led to those amazing advances. Identification of the genetic material† and its structure provided the starting point for a proper understanding of cancer, along with the ability to keep cells alive in the laboratory. Gene cloning and technology, which allowed scientists to scrutinise individual proteins picked out from the thousands of proteins in a cell,[2] led to a fundamental understanding of how signalling pathways control the cell cycle. The identification of signalling proteins and an appreciation of how they are converted by mutation into oncogenes was greatly aided by the analysis of transforming viruses and toxin-producing bacteria.

---

*As cancer is a disease that is more prevalent in older people, it becomes relatively more important as other diseases, such as infectious diseases, can be treated so that people live longer.

†The experiment that conclusively showed that nucleic acids, and not protein, contained the genes was one involving bacteria.

Aside from this virtuous role in helping to unravel cancer, evidence is now accumulating that toxins might have a more sinister role with regard to cancer. The suggestion that infectious agents can be part of the complicated causation of cancer is not new. The evidence linking several viruses to cancer is absolutely incontrovertible, particularly with regard to viruses such as the human papilloma virus, which is a major factor in cervical cancer. This virus makes proteins that interfere directly with the cell suicide programme, and vaccine development against the virus is now at an advanced stage.

The story with bacteria and cancer followed a different course. After Koch and Pasteur's seminal work showing that bacteria caused disease, some, such as the Edinburgh surgeon William Russell, proposed that all disease might be caused by bacteria. From about 1890 onwards, people noticed that bacteria clustered around tumours and suggested that these were the cause of disease. At that time, the long time lag between a triggering event and an eventual cancer was not known. In any case, these claims were soon dismissed, and the concept that bacteria could play a role in cancer was not subsequently favoured.

However, the idea was kept alive by believers who came from a different tradition to Koch and Pasteur. The origins for this group went back to an old battle that had been fought towards the end of the nineteenth century between Koch and Ferdinand Cohn on one side and Carl von Nägeli and Antoine Béchamps on the other. This latter faction believed that there were only a few types of bacteria, and that the many different bacterial types that could be observed down the microscope represented different forms of the same limited range of bacteria. One of these forms was claimed to cause cancer and the phrase 'the cancer microbe' came to be used. People who have taken this view have generally fallen out with the mainstream scientific establishment. Their views have not been published in peer-reviewed journals,* but in pamphlets or books, and now increasingly on a multitude of internet sites. The main result was that the question of whether bacteria could cause cancer fell even more into disrepute. However, everything changed in the early 1990s with the startling discoveries about stomach cancer and *Helicobacter pylori*.

---

*The process whereby an article intended for publication is reviewed anonymously by others working in the same general subject area. It is not a perfect system, but probably the least bad way to organise publication.

The evidence linking *H. pylori* to stomach cancer is strong and comes from the analysis of very large numbers of patients and controls, plus experimental infections of animals. Several molecular mechanisms appear to be involved, including the CagA protein, but a full picture of how *Helicobacter* brings about cancer is not known. Once it was accepted that one bacterium could promote cancer, it was important to know whether other bacteria had similar properties. It has already been mentioned that typhoid carriers have an increased risk of some cancers, although the mechanism involved is not known. The only other bacterial infection with a proven link to cancer is a mouse intestinal disease that is rather like human colon cancer. Other bacteria–cancer links have been suggested, but the current evidence is not strong. An important feature common to these examples is that each represents a chronic infection that will provide long-term exposure. This trait is shared with other causes of cancer, regardless of whether the inducing factor is a viral infection, exposure to radiation such as X-rays or sunlight, or a chemical carcinogen.

A completely different approach is to look for toxins with a mechanism of action that suggests that they could promote cancer. Toxins that damage DNA either directly by enzymatic attack, or indirectly by prolonged stimulation of the immune system, would be prime candidates. Similarly, toxins that interfere with the signalling pathways that regulate growth or the ability of the cell to commit suicide could encourage tumour formation.

Several toxins possess these properties. The cytolethal distending toxins (CDTs) are enzymes that cut DNA, and affected cells are known to behave as if they have suffered radiation damage. These toxins are found in several bacteria that cause gut infections. Nevertheless, at present there is no proof of a role in cancer.

Many toxins interfere directly with signalling mechanisms. The *Pasteurella multocida* toxin acts inside cells on a signalling target, the identity of which has not yet been identified, to activate several signalling pathways which strongly stimulate the growth of the target cell. Many of the signalling molecules that the toxin stimulates are known to be strongly associated with cancer. These bacteria produce a respiratory disease in pigs, so it is unlikely that this bacterium and its potent toxin are involved in human cancer. It is, however, an excellent system for studying how bacteria might cause cancer.

Another toxin that makes cells grow is the *Bacteroides fragilis* toxin—it is a protease that cleaves the cell surface molecule E-cadherin to disturb signalling mechanisms. This bacterium is a normal constituent of the gut, although it can cause diarrhoea. Its toxin appears to have the appropriate properties for involvement in cancer, but this is a totally untested hypothesis at present.

Most urinary tract infections, including cystitis and prostatitis (inflammation of the prostate gland), are caused by a strain of *E. coli* that makes the cytotoxic necrotizing factor (CNF). As mentioned in Chapter 5, CNF stimulates the important Rho proteins and in turn activates the enzyme cyclo-oxygenase 2 (COX-2), which is known to be over-expressed in many cancers. Recently, it has been shown that what were once thought to be recurrent urinary tract infections may in fact be chronic infections where the bacteria never completely go away. Women who had recurrent infections appeared always to be colonised by the same strain of *E. coli*, although this differed between patients. The only way that this could happen was if each person had been continually infected by the one strain, which had lain dormant during the time between illnesses. It was then shown, using a mouse model of infection, that the bacteria could hide from antibiotics and immune attack by living in little communities inside the bladder. If these bacteria in their dormant state release even small amounts of CNF, this could clearly influence the development of urinary tract cancers, and several researchers are beginning to suggest that this might be a possibility.

One further issue is important and relevant not just to this topic but generally to other aspects of toxin action. We normally think of toxins as molecules that kill—with the few exception of toxins such as that of *Pasteurella multocida* which makes cells grow. However, what happens in the laboratory does not always reflect what happens in life. We usually treat cells in the laboratory with large amounts of toxin, in order to produce a maximum effect, which is easiest to observe and is often more reproducible. This lets us see exactly what the toxin is capable of. If its target is a crucial signalling protein (Rho in the case of CNFs) or DNA itself (in the case of CDTs), the result for the cell may well be catastrophic. However, in an infected animal or person, different cells may be exposed to vastly different amounts of toxin. Some cells may be killed because large amounts of toxin excessively activate a crucial signalling pathway. On the other hand, cells hit by lesser amounts of toxin may survive, although this less extreme but long-term stimulation of a signalling pathway may lead to effects such as cancer. This issue has not been addressed properly for any toxin-based disease.

It therefore appears likely that toxins may not just be the main cause of damage and illness in infectious disease, but may also in some circumstances predispose towards cancer. Indeed other diseases not currently thought to be infectious may also turn out to have a bacterial origin. This would open the way for novel and effective therapies to prevent such diseases.

Finally, what of the people involved in toxins today? Are there any characters around now like the overly serious or flamboyant characters of the eighteenth and nineteenth centuries? The medically qualified people who denigrate those who are science trained—and vice versa? Those who are excessively pleased with themselves? Those who when they speak to you are constantly looking over your shoulder to see if a more important person is around to whom they should speak? Those who are ruthlessly determined to beat the competition at all cost? The big operators who try to grind the little people? And the genuine people, who are honest, generous about ideas, and care about science and truth? Do these types of people still exist, or are things different in the twenty-first century?

The answer is obvious. Human nature does not change that quickly. But of course I am not going to discuss whom these people might be. The story of the twentieth and twenty-first century toxinologists and their foibles will be another story told in another time. What is certain is that the science of these cunning but dangerous poisons and the stories of the scientists involved still have a long way to go.

Endnotes

CHAPTER 1: TOXINS ARE EVERYWHERE

1. This quotation is widely reported, including on various university sites, but the original source cannot be located. The US government website states that, when asked about it, Dr Stewart could not recall 'whether or not he made this statement'.
2. See website www.macalester.edu/~cuffel/plague2.html; see also Major (1948) in the Bibliography.
3. From First Day, Introduction (*Decameron*) (see www.brown.edu/Departments/Italian_Studies/dweb/dec_ov/dec_ov.shtml).
4. See Cartwright and Biddiss (2000).
5. Bacteria are named with a genus name, as in *Yersinia*, and a species (one member of a genus) name, as in *pestis*. The convention is to use the full name the first time that a bacterium is mentioned in any piece of writing, and then abbreviate it thereafter, as in *Y. pestis*. The names are italicised because they are Latinised names. The genus names are generally named after famous microbiologists, and the bacteria identified in the nineteenth century were often named after their discoverer. *Yersinia* is named after Yersin. There are usually several species within one family, for example the genus *Yersinia* has the species *Y. pestis*, *Y. pseudotuberculosis*, and *Y. enterocolitica*. Members of a genus are closely related, whereas different genera are more distantly related.

CHAPTER 2: THE GERM OF AN IDEA

1. See Major (1948).
2. The Royal Society was founded in 1660 to promote science. It is in effect the UK National Academy of Science, and promotes science by holding scientific meetings, as well as meetings and talks open to the public. Fellowship of the Royal Society is granted to the most eminent of UK scientists and is a high honour, shown by the initials FRS after the individual's name.
3. The range and brilliance of Hooke's work are only now beginning to be appreciated. Isaac Newton's paranoid opposition to both him and his science poisoned Hooke's reputation. The two men were different,

although each contributed in a major way to many aspects of science. The enmity between the two men was spiteful and vicious. After Hooke's death, the brilliant but difficult Newton used his powerful position as President of the Royal Society to try to erase Hooke's contributions from history.

4. Van Leeuwenhoek looked at pepper water because he hoped to discover why pepper had taste. The water had been left lying around for a few weeks, which had allowed bacteria to grow in it.

5. See Dobell (1932).

6. There would be bacteria and other forms of life in the rainwater in the bowl, although too few to be seen down the microscope. If supplied with a food source these few bacteria would grow and reproduce to produce many bacteria, as happened in the experiments on the germ theory of life discussed in this chapter and Chapter 3.

7. The first microscope is thought to have been developed in 1595 by Zacharias Jansen in the Netherlands, about 80 years before van Leeuwenhoek began making his microscopes. The earliest microscopes consisted of two lenses in a tube. The addition of a third lens later improved the design. However, the poor quality of the lenses, often made by squashing molten glass between pieces of wood, was a major drawback. The aberrations produced by one such lens are greatly multiplied in an instrument with two or more lenses.

8. See Colbatch (1721).

9. Quarantine means 40 days. It was thought that a 40-day isolation of anyone suspected of being infected would give the disease time to manifest itself. If they had not shown signs of the disease after this time, then they were not infected and could be permitted to mix with others. This practice was particularly used to try to prevent ships bringing in the plague. Although it was a good idea, not enough was known to apply it properly. Rats could scramble ashore down mooring ropes to infect the local population.

10. Puerperal means related to childbirth. Puerperal fever was a disease that probably really arose in the mid-seventeenth century, when maternity hospitals began. The advantages of having medical attention on hand were counterbalanced by the likelihood of infection, as a result of dirty instruments, the high level of infection in the hospital environment, and direct infection from doctors' hands.

11. Von Pettenkofer was appointed as the world's first Professor of Hygiene in Munich in 1859. He was not a great supporter of the germ theory of disease and later did much to oppose Robert Koch's concept of microbiology.

12. The enmity felt towards Chadwick was such that it was only close to the end of his life that his great public service was rewarded with a knighthood.

13. Karl Marx lived in this area at the time of the outbreak, whilst Florence Nightingale, famous for her work in the Crimean War and the founder of modern nursing, helped to nurse the victims of the outbreak.

14. Work linking contaminated water supplies to the incidence of typhoid was carried out by William Budd around the time of John Snow but not published until later.

CHAPTER 3: THE GOLDEN AGE OF MICROBIOLOGY

1. Photographs from the nineteenth century usually show serious poses because these are easier to hold for the long exposures needed. Nevertheless, it appears that the stern pictures of Pasteur also reflected his character. It has never been suggested that he possessed any trace of a sense of humour.

2. Normal light exists in waves that are all at different angles. Light in one plane can be selected using a polarising filter, as in polarising sunglasses.

3. Enzymes are catalysts that help a chemical reaction to take place. In chemical reactions two or more compounds interact to produce a new set of compounds. The bonds that hold the atoms together in the starting chemicals have to be broken before new bonds can be formed with atoms from the other starting compound(s). This bond-breaking process requires the input of energy. Even a potentially explosive mixture such as oxygen and hydrogen will not react until energy is supplied to begin the process of breaking the bonds between the two oxygen atoms and between the two hydrogen atoms. A catalyst lowers the amount of activation energy that is required to break the starting bonds. An enzyme does this by binding in a very specific way to the starting compounds. In this way, enzymes can dictate how any particular compound will react because they can help a chemical reaction to take place at a low temperature that would require a high temperature without the catalyst.

4. Many bacteria are anaerobic and live in what at first sight might be thought to be very oxygen-rich conditions, such as in the mouth. Such bacteria live in communities called biofilms—dental plaque being a classic example—where anaerobic bacteria can find a niche protected from oxygen.

5. Pasteur moved back to the École Normale claiming that he would rescue its fading reputation with his vigorous leadership. As at Lille, Pasteur showed his talent for administration, and it seems that he succeeded in his goal. One particular driving force was the rivalry between the École Normale and the École Polytechnique. The latter had overtaken the École Normale before Pasteur's arrival, but, perhaps of more significance, it had refused Pasteur entry as a student in 1842. Under Pasteur's directorship the École Normale's reputation dramatically improved.

6. Animal hair originating from China for use in shaving brushes was found to be badly contaminated with anthrax, and for a long time shaving brushes had by law to be sterilised before sale.

7. It is possible to make vaccines that are live and attenuated, or killed by heat or chemicals. The key issue is whether the vaccine can stimulate the immune system to remember the signature of the pathogen.

8. The vaccine against smallpox was cowpox, which is sufficiently similar to smallpox to induce protection, but it was not an attenuated vaccine, which is a version of the pathogen that has been weakened by one means or another. It is probable that the attenuation of the anthrax bacterium was caused by loss of the plasmid that holds the genes for the anthrax toxins. A plasmid is a small piece of DNA that is separate from the main circle of bacterial DNA. Plasmids often carry genes that are important for disease, such as toxins, and also proteins that code for antibiotic resistance. As they are less stable than the main bacterial DNA, they can be easily lost if the bacterium is stressed. This is a genetic change—the only difference between this change and one directly engineered in the laboratory by gene manipulation is that the first type is unpredictable and might be able to change back to produce a virulent organism.

9. There are still those who refuse to believe that there are thousands of different distinct bacteria. The existence of large numbers of bacteria on Earth can now be formally proved by obtaining the DNA sequence, which is the current way that bacteria are grouped into species. However, Béchamps and Nägeli were followed in the twentieth century by the German scientist, Günther Enderlein, who was championed by Karl Windstosser. They developed Béchamps' ideas and came up with a whole new system of obscure terminology for bacterial forms. Not surprisingly, Enderlein was shunned by the mainstream scientific and medical community. One of Enderlein's disciples, Royal Rife, believed there to be only about 10 different bacteria. He claimed that *Escherichia coli* could change into *Salmonella typhi* (the agent of typhoid), then yeast and then viral forms, one of which he further claimed could be isolated from all cancerous tumours. Although there is now evidence to link some bacteria to

cancer (discussed in Chapter 9), current theories owe nothing to Enderlein's eccentric ideas, which serve only to confuse.

10. Fanny Hesse was born in New Jersey, but had net and married Walter Hesse on her European tour. She had heard about agar from East Indian friends of her mother. Koch had previously used gelatin, but this melted at body temperature.

11. Metchnikoff was a colourful character. Pasteur's son-in-law Valery-Radot described how Metchnikoff showed an interest in one of Madam Pasteur's young housemaids, asking if she was a virgin. When she assured him that she was, he expressed his delight, claiming that he had been looking for a virgin since he had come to Paris. He was interested in studying the bacteria growing in a virginal vagina. However, Valery-Radot points out that Metchnikoff 'studied that flora so well that he made her pregnant'.

12. Public knowledge about the rabies vaccine, which was made known by Pasteur, does not correspond exactly with some of the problems and ethical difficulties revealed by subsequent analysis of Pasteur's laboratory notebooks. Rabies is a viral disease and the concept of a virus was totally unknown until years later. Furthermore, the effectiveness of Pasteur's treatment with live attenuated rabies vaccines was, by the nature of the disease, very difficult to determine. It is interesting that, at the Institut Pasteur, Pasteur's rabies vaccine was soon replaced by one that had been chemically inactivated.

13. Occasionally shown in the UK, this is Louis Pasteur the Hollywood way. Events and places are changed or invented, Koch is ignored, and a romantic interlude is imposed. Yet it still manages to get across some of the science in an interesting way.

14. Koch's postulates can be stated simply as:

   1. The bacterium must be isolated from diseased people and not from healthy people. (This shows that it has a causative link with disease.)

   2. The location of the bacterium in the body should correspond to the location of the disease signs. (For example, you would expect to find bacteria in the gut for a gut disease such as diarrhoea.)

   3. After isolating the bacterium, growing it in the laboratory and giving it to experimental animals, it should cause the same disease in these animals. (Thus the bacterium, uncontaminated by other pathogens, causes the disease.)

15. *Bacillus anthracis*, the cause of anthrax, is a large bacterium that is easy to see in the microscope, and grows relatively quickly. In contrast, *Mycobacterium tuberculosis* is small, does not stain easily, and grows incredibly slowly.

16. The ageing Max von Pettenkofer, the Professor of Hygiene in Munich, drank a flask of cholera bacteria and suffered only mild symptoms, which he insisted were not caused by cholera. It has been suggested that Gaffky, who had sent the culture to von Pettenkofer, had guessed the reason for the request and intentionally sent a weak strain of the bacterium.

17. See Ogawa (2000).

18. This institute was set up by the British in 1889 in Pune and transferred to Mukteswar in 1893. It was later renamed the Indian Veterinary Research Institute, and later still the Mukteswar laboratory became a field station of a larger institute.

19. Ehrlich later made many contributions, including pioneering the use of chemicals to treat diseases, notably salvarsan as a cure for syphilis, and thus founded chemotherapy—the treatment of disease with chemicals.

20. Gram-positive bacteria differ from Gram-negative bacteria in the structure of the cell wall that surrounds the bacteria. In Gram-positive bacteria there is a very thick layer of molecules made from sugars and proteins, called peptidoglycan. This holds the stain inside the bacterium. The Gram-negative cell has a much thinner layer of peptidoglycan and the stain is easily washed out.

21. Immunology is the study of the immune system, that is the way the body recognises and deals with foreign invading organisms. Exposure to an infectious agent, or a vaccine that partially mimics the disease, primes the body so that it is able to recognise as foreign a pathogen should it encounter it a second time.

CHAPTER 4: THE ANATOMY OF DIPHTHERIA

1. Damage to the body from diphtheria can take many forms. The Spanish composer Joaquin Rodrigo, best known for his lyrical Concierto de Aranjuez for guitar and orchestra, was blind from the age of three after contracting diphtheria.

3. See Major (1948).

3. Bacteria make some proteins only when they judge that the conditions are right. The artificial food in the laboratory was not close enough to life in a throat, until the bacteria were allowed to remain in the broth for longer.

4. Pappenheimer was known with some affection by friends and colleagues as 'Pap', an interesting nickname hinting at the fatherly role that many have ascribed to him. In his tribute to Pappenheimer, John Collier describes a man with broad interests, married for almost 60 years, father

of three children, a keen clarinettist and a viola player, who promoted music concerts during his time at Harvard, and who was fond of socialising—to the extent of skinny dipping 'preferably in the company of members of the opposite sex'.

5. MacLeod was a key figure in the DNA story. Along with Oswald Avery and Maclyn McCarthy he showed conclusively in 1944 that genes were made of DNA and not, as was previously believed, protein. This was a seminal result and one that was essential for the later explosion of work on DNA.

6. Ribosomes are large structures made of protein and RNA (ribosomal RNA or rRNA) that act as a one-way dictionary to translate the message of the genes into protein. Ribosomes are complex and interact with several types of molecule. They have to interact with the messenger RNA (mRNA) that carries a copy of the gene from the nucleus, where the DNA is, to the main part of the cell (the cytoplasm), where the ribosome is. Transfer RNAs (tRNAs) form a family of adapter molecules, each of which is linked to a particular amino acid. Each type of tRNA carries a particular amino acid to the ribosome. There, if its recognition sequence interacts with the related three-letter code of the mRNA, its attached amino acid is added to the growing chain of protein.

7. Hydrophobic means hating water. Hydrophobic molecules, such as fat molecules, are electrically neutral and interact poorly with water, but well with similar molecules. Hydrophilic means water loving, and implies that the molecule is either electrically charged or has an uneven electrical charge distribution, that is a polar distribution with part of the molecule slightly positive and some of it slightly negative. Water is polar because its oxygen atom attracts the electrons more strongly than hydrogen, so that that end of the molecule is slightly negatively charged whereas the hydrogen end is slightly positively charged. This accounts for many of the properties of water.

8. This three-dimensional picture was obtained by bombarding solid crystals of diphtheria toxin with X-rays and analysing how the X-rays were scattered. This sort of analysis shows where every atom is, and thus where every amino acid is, and importantly how they are grouped together.

CHAPTER 5: UNDERSTANDING

1. Whether the ring forms before or at the time of insertion probably differs from toxin to toxin.

2. This process is called differentiation. It starts with the original fertilised egg that has just received a sperm. This is an omnipotential cell, a stem cell that has the potential to make all types of cell. As the egg grows and develops in the womb, cells lose this ability. They follow a pathway of differentiation that eventually leads to a fully differentiated cell which does not have the ability to form other cell types and is highly specialised, such as a nerve cell or bone cell.

3. Radiation and certain chemicals injure DNA molecules, leading to changes in the bases that encode the genes. In a resting cell this damage can often be repaired. In a fast-growing cell that is rapidly copying its DNA for new cells, there may not be time to repair the damage. Thus the changed DNA may be copied so that the change is fixed into the DNA of the new cells. If the treatment has damaged a key gene the cells will die. Radiation and some chemicals can also, in other circumstances, cause cancer by damaging DNA in previously healthy cells.

4. The first chemical cause of cancer was recognised in the eighteenth century by Percival Potts. The high rate of cancer of the scrotum in young chimney sweeps was linked to their exposure to tarry compounds as they climbed through the chimneys to clean them. The link of radiation with cancer occurred shortly after its discovery.

5. There are around 20 of these G-proteins. The four classes are G_s (stimulatory), G_i (inhibitory), G_q (isolated using Q columns—a type of resin that binds proteins), and G_{12} (G-protein number 12). They are also called the heterotrimeric G-proteins, because they are made up of three (trimeric) different (hetero-) protein chains.

6. The receptor is called a toll-like receptor because it is the human equivalent of the toll receptors found in the fruit fly. *Toll* is German slang for 'cool', which was allegedly said when Christiane Nusslein-Volhard saw evidence for these receptors in 1980—work that in part later lead to a Nobel Prize for her in 1995.

7. Pfeiffer was born in 1858 in present-day Poland. His success with endotoxin led to his appointment as Professor of Hygiene in Königsberg in 1899 and later in Breslau, where Koch had visited Ferdinand Cohn some 25 years earlier to convince him of his discoveries with anthrax.

8. It is now known that, as well as making endotoxin, *V. cholerae* makes other toxins. In particular it releases cholera toxin that directly causes the watery diarrhoea. Cholera toxin is a protein toxin that is destroyed by heat.

9. Protective antigen is so called because it is the basis of the anthrax vaccine. The immune response that immunised animals make against this protein prevents it from carrying out its normal function of helping the two toxic factors into the cell to cause damage.

10. CDT is produced by *Shigella*, *E. coli*, *Actinobacillus actinomycetemcomitans* (one cause of periodontal disease), *Haemophilus ducreyi* (cause of the sexually transmitted infection, chancre), *Helicobacter hepaticus*, and *S. typhi* (the cause of typhoid).

CHAPTER 6: WHY ARE PLAGUE AND TYPHOID SO DEADLY?

1. *S. typhimurium* causes a typhoid-like disease in mice (*murium* is from the Latin for mouse). *S. typhimurium* has greatly helped our understanding of all *Salmonella* species, including those that cause gastrointestinal disease.

2. HeLa cells were the cells that had helped Alwin Max Pappenheimer and his colleagues to understand diphtheria. They were isolated from a cervical cancer and were among the first cells to be grown in the laboratory.

3. Compare this with the 150 million years since it is thought that *Escherichia coli* became a separate species from *Salmonella*. *E. coli* and the *Salmonella* group of bacteria are viewed as being remarkably close cousins.

4. Syphilis, when it first appeared in the west in the late fifteenth century, was apparently more virulent than now. Similarly the terrible pandemic of influenza in 1918, although not a new disease, produced a different disease pattern in terms of its severity than seen since, for reasons that are still unknown.

5. Ebola is a deadly haemorrhagic virus (virus that causes haemorrhage or bleeding) that leads to massive internal bleeding and a high death rate. It has caused several explosive, but thankfully limited, outbreaks in Africa.

CHAPTER 7: DEVIANT BIOLOGY

1. Herman Stillmark identified ricin in Germany in 1889. The toxin was extensively used by Paul Ehrlich in studies that put immunology on a firm base. Given orally, in the form of seed, ricin is 1,000 times less toxic than when injected or inhaled, and it gives rise to antibodies. Antibodies were identified by Ehrlich as the body's response to this toxic protein. Ricin

was later shown to be an AB toxin that attacks the large RNA molecule in the ribosome to inhibit protein synthesis.

2. As well as biological experiments, there were tests with chemical weapons. Other macabre investigations were carried out. People were hung upside down to see how long they could survive, some were subjected to high or low pressures so that their organs burst, and others were staked out on the ground in the heat to see how long they could survive without water. The frostbite experiments consisted of freezing a limb, trying to treat it, cutting it off, and then starting on another limb.

3. The suggestion that American soldiers were used in biological experiments broke publicly in 1976 following a Japanese documentary. Two Congressional hearings in the 1980s heard veterans' testimony about their treatment by the Japanese, but failed to produce a report, despite some astounding revelations. A retired senior army archivist admitted that boxes of information seized at the end of the war had been sent back to the Japanese government, without a copy being kept—because it would have been too difficult to have the papers translated. A 1995 newspaper report suggested that some Americans who survived being shot down in 1945 in Fukuoka were taken to nearby Kyushu University where they were experimented on. The story leaked out and several people were tried and found guilty at the Yokohama war crimes trial, although the sentences were never carried out. Ishii himself admitted to using Russian prisoners, but always denied that Americans had been experimented on.

4. Some archaeologists excavating an old hospital at Soutra, just south of Edinburgh, claimed to have found evidence of anthrax spores, even though the hospital had been closed for over 500 years. Unfortunately none of the details of the Soutra excavation appears to have been published, except as six reports and in conference proceedings that have been lodged in the National Library of Scotland. These reports, deposited on behalf of the Soutra Hospital Archaeoethnopharmacological Research Project, and quaintly referred to by the authors as SHARP Practice 1–6, give no substantiated details about the microbiological findings, so the accuracy of the claims must be questioned.

5. Against this cautious view, it has to be remembered that, during the first tests of both fission and fusion nuclear weapons, it was not known whether they would lead to the annihilation of the human race. However, the tests went ahead.

6. Much of the information in this and the next section came from a detailed on-line document written and continually updated by Dr Seth Carus at the National Defense University in Washington DC.

7. As many of the senior members of the cult were women they were referred to as the 'moms', and the more senior as the 'big moms'. Sheela was a very powerful figure, who either discussed decisions with the Bhagwan or took them herself. The cult operated by threatening legal action to those who opposed it.

8. It is known that botulinum toxin can be effective as an aerosol. In the 1960s three laboratory workers became ill several days after performing a postmortem examination on an experimental animal. They were treated with anti-toxin serum and all recovered.

CHAPTER 8: A MORE OPTIMISTIC OUTCOME

1. The re-use of needles has been shown to transmit AIDS and hepatitis, and the World Health Organization (WHO) has expressed concerns about the adequate sterilisation of syringes. There is the added problem of the safe disposal of used needles and syringes.

2. The girl was Bessie Dashiell and her claim to fame was her friendship with the shy young John D. Rockefeller Jr, soon to become one of the wealthiest Americans ever, as heir to the Standard Oil millions. Her death had a devastating effect on the teenage Rockefeller, and is thought to have encouraged his great philanthropic works, the Rockefeller University and the Memorial Sloan-Kettering Cancer Center in New York.

3. The great advances over the last 30 years in understanding how the cell worked and what went wrong in cancer showed that certain cell surface proteins were over-expressed in cancer cells.

4. A Florida doctor has recently been reported to have injected unlicensed botulinum toxin of around 100,000 times the strength of BOTOX® into himself and some of his patients. They are currently paralysed and it is not known whether they will make a full recovery. There are also reports of black market botulinum toxin A, mostly from China.

5. I am not sure when this term was first used, but a nice review that covers this topic by Giampietro Schiavo and Gisou van der Goot, published in 2001, used 'toolkit' in its title (see Bibliography).

6. GM1 is a component of lipid rafts, which, as the name implies, are patches of membrane that differ from other parts of the membrane. Rafts are thought to be very important in almost all aspects of membrane function. In other words, the membrane of a cell is a highly ordered arrangement of different proteins and lipids, where different parts of the membrane perform distinct functions.

CHAPTER 9: WHERE IS TOXINOLOGY GOING NOW?

1. Toxinology is the science of toxins. It is a word favoured by Joseph Alouf, the genial and fatherly co-founder of the European Toxin meetings, a biannual 'toxinfest' that is the premier international forum for discussing toxin science. The most recent meeting was held in Canterbury, England in June 2005. Toxinology is not to be confused with toxicology—the study of toxic substances in the body.

2. Antibodies made against foreign invaders are so specific that they attach only to one type of protein and so can pick out that protein. The artificial production of antibodies against a particular protein can then be used to identify that protein experimentally. Other techniques enable the analysis of all the proteins in a cell, an approach called proteomics.

Bibliography

Books

Ackerknecht, E.H. (1982) *A Short History of Medicine*. John Hopkins University Press, London.

Acland, H.W. (1856) *Memoir of the Cholera at Oxford in the Year 1854*. John Churchill, London.

Alouf, J.E. and Freer, J.H. (eds) (1991) *Sourcebook of Bacterial Protein Toxins*. Academic Press, London.

Beck, R.W. (2000) *A Chronology of Microbiology in Historical Context*. American Society for Microbiology Press, Washington DC.

Brock, T.D. (1999) *Robert Koch*. American Society for Microbiology, Washington DC.

Bruel, W. (1632) *The Physician's Practice*. John Norton, London.

Cartwright, F.F. and Biddiss, M. (2000) *Disease and History*. Sutton Publishing Ltd, Stroud.

Colbatch, J. (1721) *A scheme for proper methods to be taken, fhould it pleafe GOD to vifit us with the PLAGUE*. J. Darby, London.

de Kruif, P. (1927) *Microbe Hunters*. Jonathan Cape, London.

Defoe, D. (1660) *Journal of the Plague Year*. Dent, London, 1966.

Dobell, C. (1932) *Antony van Leeuwenhoek and His 'Little Animals'*. John Bale, Sons & Danielsson Ltd, London.

Dormandy, T. (2003) *Moments of Truth*. John Wiley & Sons, Chichester.

Geison, G.L. (1985) *The Private Science of Louis Pasteur*. Princeton University Press, Princeton, NJ.

Hall, S.S. (1999) *Commotion in the Blood: Life, Death and the Immune system*. Diane Publishing Co.

Harris, S.H. (1994) *Factories of Death*. Routledge, London.

Holmes, S.J. (1924) *Louis Pasteur*. Harcourt, Brace & Co., London.

Karlen, A. (1995) *Plague's Progress*. Victor Golancz, London.

Lax, A.J. (ed.) (2005) *Bacterial Protein Toxins that Interfere with the Regulation of Growth*. Cambridge University Press, Cambridge.

Longmate, N. (1970) *Alive and Well*. Penguin Books, Harmondsworth.

Major, R.H. (1948) *Classic Descriptions of Disease*. Blackwell Scientific Publications, Oxford.

Mead, R. (1720) *A Short Discourse Concerning Pestilential Contagion, and the Methods to be Used to Prevent It*. Samuel Buckley and Ralph Smith, London.

Pye, G. (1721) *A Discourse of the Plague; wherein Dr Mead's notions are considerd and refuted.* J. Darby, London.

Robinson, V. (1935) *The Story of Medicine.* Tudor Publishing Co., New York.

Scott, S. and Duncan, C.J. (2001) *Biology of Plagues: Evidence from Historical Populations.* Cambridge University Press, Cambridge.

Sidell, F.R., Takafuji, E.T. and Franz, D.R. (1997) *Textbook of Military Medicine: Medical Aspects of Chemical and Biological Warfare.* Office of the Surgeon General at TMM Publications, Borden Institute, Walter Reed Army Medical Center, Washington DC.

Tanaka, T. (1996) *Hidden Horrors: Japanese War Crimes in World War II.* Westview Press, Oxford.

Vallery-Radot, R. (1960) *The Life of Pasteur.* Dover Publications, New York.

Ziegler, P. (1969) *The Black Death.* Collins, London.

Scientific papers

Achtman, M., Zurth, K., Morelli, G., Torrea, G., Guiyoule, A. and Carniel, E. (1999) *Yersinia pestis,* the cause of plague, is a recently emerged clone of *Yersinia pseudotuberculosis. Proceedings of the National Academy of Sciences of the USA* **96,** 14043–14048.

Ben-Yehuda, A., Belostotsky, R., Lichtenstein, M., Aqeilan, R., Abady, R. and Lorberboum-Galski, H. (2002) Chimeric proteins as candidates for cancer treatment. *Drugs of the Future* **27,** 1–10.

Bickels, J., Kollender, Y., Merinsky, O. and Meller, I. (2002) Coley's toxin: historical perspective. *Israeli Medical Association Journal* **4,** 471–472.

Bonn, D. (2000) *Clostridium novyi* revealed as heroin contaminant. *Lancet* **355,** 2227–2230.

Cohen, J. (2002) The immunopathogenesis of sepsis. *Nature* **420,** 885–891.

Collier, R.J. (1996) A tribute to Pap. In: Frandsen, P.L., Alouf, J.E., Falmagne, P. et al. (eds), *Bacterial Protein Toxins.* Gustav Fischer Verlag, New York, pp. 4–7.

—— (2001) Understanding the mode of action of diphtheria toxin: a perspective on progress during the 20th century. *Toxicon* **39,** 1793–1803.

Cornelis, G.R. and Wolf-Watz, H. (1997) The *Yersinia* Yop virulon: a bacterial system for subverting eukaryotic cells. *Molecular Microbiology* **23,** 861–867.

Crumpton, R. and Gall, D. (1980) Georgi Markov—death in a pellet. *Medico-Legal Journal* **48,** 51–62.

Dennis, D.T., Inglesby, T.V., Henderson, D.A. et al. (2001) Tularemia as a biological weapon. Medical and public health management. *Journal of the American Medical Association* **285,** 2763–2773.

Dixon, T.C., Meselson, M., Guillemin, J. and Hanna, P.C. (1999) Anthrax. *New England Journal of Medicine* **341,** 815–826.

Drancourt, M. and Raoult, D. (2002) Molecular insights into the history of plague. *Microbes and Infection* 4, 105–109.

Elworthy, R.R. (1916) An outbreak of anthrax conveyed by infected shaving brushes. Lancet i, 20–23.

Erbguth, F.J. and Naumann, M. (1999) Historical aspects of botulinum toxin: Justinus Kerner (1786–1862) and the 'sausage poison'. *Neurology* 53, 1850–1853.

Foran, P.G., Mohammed, N., Lisk, G.O. et al. (2003) Evaluation of the therapeutic usefulness of botulinum neurotoxin B, C1, E, and F compared with the long lasting type A. Basis for distinct durations of inhibition of exocytosis in central neurons. *Journal of Biological Chemistry* 278, 1363–1371.

Gest, H. (2004) The discovery of microorganisms by Robert Hooke and Antoni van Leeuwenhoek, fellows of the Royal Society. *Notes and Records of the Royal Society of London* 58, 187–201.

Gronski, P., Seiler, F.R. and Schwick, H.G. (1991) Discovery of antitoxins and development of antibody preparations for clinical uses from 1890 to 1990. *Molecular Immunology* 28, 1321–1332.

Gross, L. (1995) How the plague bacillus and its transmission through fleas were discovered: reminiscences from my years at the Pasteur Institute in Paris. *Proceedings of the National Academy of Sciences of the USA* 92, 7609–7611.

Hinnebusch, B.J., Rudolph, A.E., Cherepanov, P., Dixon, J.E., Schwan, T.G. and Forsberg, Å. (2002) Role of *Yersinia* murine toxin in survival of *Yersinia pestis* in the midgut of the flea vector. *Science* 296, 733–735.

Izumi, Y. and Isozumi, K. (2001) Modern Japanese medical history and the European influence. *Keio Journal of Medicine* 50, 91–99.

Lawrence, H.S. (2000) Alwin Max Pappenheimer Jr. *Biographical Memoirs* 77, 265–280.

Ludwig, A. (1996) Cytolytic toxins from Gram-negative bacteria. *Microbiología* 12, 281–296.

Manchee, R.J., Broster, M.G., Melling, J., Henstridge, R.M. and Stagg, A.J. (1981) *Bacillus anthracis* on Gruinard Island. *Nature* 294, 254–255.

—— —— Stagg, A.J. and Hibbs, S.E. (1983) Formaldehyde solution effectively inactivates spores of *Bacillus anthracis* on the Scottish Island of Gruinard. *Applied and Environmental Microbiology* 60, 4167–4171.

Manning, G., Whyte D. B., Martinez R., Hunter T. and Sudarsanam S. (2002) The protein kinase complement of the human genome. *Science* 298, 1912–1934.

Moffat, B. (1988) An investigation of medieval medical treatments: Soutra Hospital archaoethnopharmacological research project (SHARP). *Royal College of Physicians of Edinburgh Proceedings* 7, 80–86.

—— (ed.) (1998) SHARP Practice 6: the sixth report on researches into the medieval hospital at Soutra Scottish Borders/Lothian Scotland. Sharp, Lothian.

Nataro, J.P. and Kaper, J.B. (1998) Diarrheagenic *Escherichia coli. Clinical Microbiology Reviews* 11, 142–201.

Ogawa, M. (2000) Uneasy bedfellows: science and politics in the refutation of Koch's bacterial theory of cholera. *Bulletin of the History of Medicine* 74, 671–707.

Pallen, M.J., Lam, A.C., Loman, N.J. and McBride, A. (2001) An abundance of bacterial ADP-ribosyltransferases—implications for the origin of exotoxins and their human homologues. *Trends in Microbiology* 9, 302– 307.

Pappenheimer, A.M. (1937) Diphtheria toxin: isolation and characterization of a toxic protein from *Corynebacterium diphtheriae* filtrates. *Journal of Biological Chemistry* 120, 543–553.

—— (1977) Diphtheria toxin. *Annual Review of Biochemistry* 46, 69–94.

Peck Gossel, P. (2000) Pasteur, Koch and American bacteriology. *History and Philosophy of Life Sciences* 22, 81–100.

Reader, I. (2002) Spectres and shadows: Aum Shinrikyô and the road to Megiddo. *Terrorism and Political Violence* 14, 147–189.

Revell, P.A. and Miller, V.L. (2001) *Yersinia* virulence: more than a plasmid. *FEMS Microbiology Reviews* 205, 159–164.

Rietschel, E.T. and Cavaillon, J-M. (2002) Endotoxin and anti-endotoxin. The contributions of the schools of Koch and Pasteur: Life, milestone-experiments and concepts of Richard Pfeiffer (Berlin) and Alexandre Besredka (Paris). *Journal of Endotoxin Research* 8, 3–16, 71–82.

Rossetto, O., Seveso, M., Caccin, P., Schiavo, G. and Montecucco, C. (2001) Tetanus and botulinum neurotoxins: turning bad guys into good by research. *Toxicon* 39, 27–41.

Russello, S.V. and Shore, S.K. (2004) SRC in human carcinogenesis. *Frontiers in Bioscience* 9, 139–144.

Schiavo, G. and van der Goot, F.G. (2001) The bacterial toxin toolkit. *Nature Reviews Molecular Cell Biology* 2, 530–536.

Smith, H. (2000) Discovery of the anthrax toxin: the beginning of in vivo studies on pathogenic bacteria. *Trends in Microbiology* 8, 199–200.

Solomon, T. (1995) Alexandre Yersin and the plague bacillus. *Journal of Tropical Medicine and Hygiene* 98, 209–212.

Soper, G.A. (1919) Typhoid Mary. *The Military Surgeon* 45, 2–15.

Strauss, N. and Hendee, E.D. (1959) The effect of diphtheria toxin on the metabolism of HeLa cells. *Journal of Experimental Medicine* 109, 145–163.

Summers, W.C. (2000) History of microbiology. In: Lederberg, J. (ed.), *Encyclopedia of Microbiology*, 2nd edn. Academic Press, London, pp. 677– 697.

Thurston, A.J. (2000) Of blood, inflammation and gunshot wounds: the history of the control of sepsis. *Australia and New Zealand Journal of Surgery* 70, 855–861.

Turnbull, P.C.B. (1991) Anthrax vaccines: past, present and future. *Vaccine* 9, 533–539.

Vitek, C.R., Brennan, M.B., Gotway, C.A. et al. (1999). Risk of diphtheria among schoolchildren in the Russian Federation in relation to time since last vaccination. *Lancet* 353, 355–358.

von Behring, E.A. and Kitasato, S. (1890) Über das Zustandekommen der Diphtherie-Immunität und der Tetanus-Immunität bei Thieren. *Deutsche Medicinische Wochenschrift* 49: 1113–1114.

von Goethe, J.W. (1808) *Faust I* (Studierzimmer; Mephistopheles).

Wallis, T.S. and Galyov, E.E. (2000) Molecular basis of *Salmonella*-induced enteritis. *Molecular Microbiology* 36, 997–1005.

Watts, J. (2002) Court forces Japan to admit to dark past of bioweapons programme. *Lancet* 360, 628, 857.

Wu, S.G., Morin, P.J., Maouyo, D. and Sears, C. (2003) *Bacteroides fragilis* enterotoxin induces *c-myc* expression and cellular proliferation. *Gastroenterology* 124, 392–400.

Scientific society magazines

Finkelstein, R.A. (2000) Personal reflections on cholera: the impact of serendipity. *American Society for Microbiology News* 66, 663–667.

Hesse, W. (1992) Walther and Angelina Hesse—early contributors to bacteriology. *American Society for Microbiology News* 58, 425–428.

Howard, D.H. (1982) Friedrich Loeffler and his history of bacteriology. *American Society for Microbiology News* 48, 297–302.

Lutzker, E. and Jochnowitz, C. (1987) Waldemar Haffkine: pioneer of cholera vaccine. *American Society for Microbiology News* 53, 366–369.

Mortimer, P. (2001) Koch's colonies and the culinary contribution of Fanny Hesse. *Microbiology Today* 28, 136–137.

Perry, R.D. (2003) A plague of fleas—survival and transmission of *Yersinia pestis*. *American Society for Microbiology News* 69, 336–340.

Stropnik, Z.C. (2002) Plenčič (1705–1786): Might we call him a microbiologist? *FEMS Circular* 52, 7.

Various authors (2004) *The Biochemist* 26, No 2.

Newspaper articles

Easton, T. (1995) Japan admits dissecting WWII POWs. *Baltimore Sun* 28 May.

Rosie, G. (2001) UK planned to wipe out Germany with anthrax. October 14 *The Herald*

Websites

http://earlyamerica.com/review/2000_fall/1832_cholera.html
http://history1900s.about.com/library/weekly/aa062900a.htm
http://nobelprize.org/
www.aiipowmia.com/731/731caveat.html
www.bordeninstitute.army.mil/ethicsbook_files/Ethics2/Ethics-ch-16.pdf
www.brown.edu/Departments/Italian_Studies/dweb/dec_ov/dec_ov.shtml
www.cdc.gov/ncidod/EID/vol5no4/kortepeter.htm
www.cf.ac.uk/hisar/teach/history/projects/plague/
www.eyammuseum.demon.co.uk/
www.gwu.edu/~nsarchiv/NSAEBB/NSAEBB61/
www.insecta-inspecta.com/fleas/bdeath/bdeath.html
www.johnsnowsociety.org/
www.macalester.edu/~cuffel/plague2.htm
www.microscopy.fsu.edu/primer/museum/index.html
www.mja.com.au/public/issues/177_11_021202/dec10354_fm.pdf
www.nbc-med.org/SiteContent/HomePage/WhatsNew/MedAspects/
    contents.html
www.ndu.edu/centercounter/prolif_publications.htm
www.pasteur.fr/
www.pasteur.fr/english.html
www.ph.ucla.edu/epi/bioter/detect/antdetect_intro.html
www.ph.ucla.edu/epi/snow.html
www.sciences.demon.co.uk/wavintr.htm
www.scotland.gov.uk/library/documents-w4/pgr-00.htm
www.sjwar.org/
www.stevenlehrer.com/explorers
www.ucmp.berkeley.edu/history/leeuwenhoek.html
www.vnh.org/MedAspChemBioWar/
www.wellcome.ac.uk
www.who.int/en/
www.wolverhampton.gov.uk/leisure_culture/culture/archives/
    publications.htm
www.wood.army.mil/chmdsd/Army_Chemical_Review/pdfs/2004%20Oct/
    MJHayesTribute-04-2.pdf
www.worldortho.com/huckstep/index.html

Further reading

Books

I have listed just a few books in each category, starting with those that appear to be most readable.

History of microbiology

De Kruif, P. (1927) *Microbe Hunters*. Jonathan Cape, London.

One of the most unusually written books ever written. Almost impossible to put it down, it oozes enthusiasm for the heroic microbe hunters, imagining their everyday conversations as they invented microbiology. It takes many liberties and has irritated many reviewers, although this book did a lot to popularise this topic. It sold over a million copies and was translated into many languages. De Kruif was trained as a bacteriologist, but was sacked from his academic position for writing a book poking fun at the medical profession.

Cartwright, F.F. and Biddiss, M. (2000) *Disease and History*. Sutton Publishing Ltd, Stroud.

Selected diseases that changed the course of history. Written by one historian and a medically qualified doctor, this is very readable.

Ackerknecht, E.H. (1982) *A Short History of Medicine*. Johns Hopkins University Press, London.

All aspects of medicine from the beginning to the start of the twentieth century, very much from a medical point of view.

Longmate, N. (1970) *Alive and Well*. Penguin Books, Harmondsworth.

An interesting little book written from the public health angle.

The plague

Karlen, A. (1995) *Plague's Progress*. Victor Golancz, London.

A well-written description of many plagues from the earliest epidemics through true plague to more modern plagues such as AIDS.

Ziegler, P. (1969) *The Black Death*. Collins, London.

A comprehensive account of the Black Death in Europe, which also details the social and economic consequences.

People

Holmes, S.J. (1924) *Louis Pasteur*. Harcourt, Brace & Co., London.
A short and readable account that discusses Pasteur's life and work in the context of previous work and his contemporaries, such as Robert Koch.
Brock, T.D. (1999) *Robert Koch*. American Society for Microbiology, Washington DC.
The first English language biography of Koch is a thorough and balanced view of all aspects of Koch, and was the major source of material about Koch used in this book.
Dormandy, T. (2003) *Moments of Truth*. John Wiley & Sons, Chichester.
Four biographies in one: Semmelweis and Lister are of relevance here.
Vallery-Radot, R. (1960) *The Life of Pasteur*. Dover Publications, New York.
Written in 1901 by Pasteur's son-in-law, this is the most comprehensive study of Pasteur, arranged in a strictly chronological order, but it is overly reverential.
Dobell, C. (1932) *Antony van Leeuwenhoek and his 'Little Animals'*. John Bale, Sons & Danielsson Ltd, London.
A scholarly and fascinating book which highlights how immense van Leeuwenhoek's contribution was. Dobell translated some of van Leeuwenhoek's seventeenth century letters to the Royal Society for the first time. Brian Ford has stated that this work was almost an obsession for Dobell, at a time that the history of science was not viewed highly.

Biological warfare

Harris, S.H. (1994) *Factories of Death*. Routledge, London.
Describes all aspects of the Japanese biological weapons programme from its beginnings, through its operation to the post war cover-up.
Tanaka, T. (1996) Hidden Horrors: Japanese War Crimes in World War II. Westview Press, Oxford.
One chapter is devoted to Ishii's Unit 731.

Websites

Websites are notoriously ephemeral but these were available at the time of publication.

History of microbiology

www.stevenlehrer.com/explorers
Now apparently out of print, Steven Lehrer's 1979 book *Explorers of the Body* originally published by Doubleday can be downloaded in its entirety from Stephen Lehrer's website. This book covers a wide range of medical

history including the rise of microbiology and describes the characters as well as their discoveries.

www.mja.com.au/public/issues/177_11_021202/dec10354_fm.pdf
    An interesting article on the history of puerperal fever, and the nineteenth-century doctors who tried to promote ideas of cleanliness to their colleagues.

www.sciences.demon.co.uk/wavintr.htm
    An extensive site dealing with van Leeuwenhoek. Brian Ford is a freelance scientist and broadcaster.

www.microscopy.fsu.edu/primer/museum/index.html
    Everything you wanted to know about microscope history and the images that microscopes can produce. An absorbing and well-illustrated site, with lots of good pictures.

Plague

www.eyammuseum.demon.co.uk
    The website of the Eyam museum that describes the village's self-imposed quarantine.

www.brown.edu/Departments/Italian_Studies/dweb/dec_ov/dec_ov.shtml
    A link to Boccaccio's *Decameron.*

www.insecta-inspecta.com/fleas/bdeath/bdeath.html
    Lots of information on the Black Death.

www.cf.ac.uk/hisar/teach/history/projects/plague/
    The historical aspects of the Black Death in England.

Cholera

www.ph.ucla.edu/epi/snow.html
    This website was set up by the Ralph Frerichs of the UCLA Epidemiology Department in California and contains many details about John Snow. It is very readable, with many fascinating snippets of information about Snow, downloadable maps, pictures, and his important pamphlets and lectures. It is well worth investigating.

www.johnsnowsociety.org/
    The website of the John Snow Society also has many interesting facts about Snow.

http://earlyamerica.com/review/2000_fall/1832_cholera.html
    An article describing how the 1832 cholera epidemic affected New York.

Typhoid

http://history1900s.about.com/library/weekly/aa062900a.htm
    One of several sites that describe the story of Mary Mallon ('Typhoid Mary').

www.worldortho.com/huckstep/index.html
> A website with a description of some of the history of typhoid and also some of the signs and symptoms of the disease.

People

www.pasteur.fr/
www.paris.org/Musees/Pasteur/info.html
> The website of the Pasteur Institute and also visitor information about the Pasteur museum based in the house where he lived. The museum is a 'must see' for any visitor to Paris, but note that it is closed in August. Pasteur's own paintings are hung on the walls, and some of his equipment, such as the swan neck flasks, is on display. His tomb is in the basement.

Biological warfare

The first two sites on Unit 731 and related Japanese atrocities contain disturbing images and articles:
www.aiipowmia.com/731/731caveat.html
> Entry to the Unit 731 pages of the 'Advocacy and Intelligence Index For Prisoners of War-Missing in Action'.
www.sjwar.org/
> Website of the 'Alliance for Preserving the Truth of Sino-Japanese War'.
www.bordeninstitute.army.mil/ethicsbook_files/Ethics2/Ethics-ch-16.pdf
> Chapter written by Sheldon Harris detailing the Japanese atrocities.
www.nbc-med.org/SiteContent/HomePage/WhatsNew/MedAspects/contents.html
> Detailed and informative US military site with a lot of information.
www.cdc.gov/ncidod/EID/vol5no4/kortepeter.htm
> An article on potential biological weapons.
www.ndu.edu/centercounter/prolif_publications.htm
> The publications website of the National Defense University in the USA. Dr Seth Carus's document provides a wealth of information on the use of biological weapons.
www.ph.ucla.edu/epi/bioter/detect/antdetect_intro.html
> Authoritative site on the American anthrax attack in 2001.

Up-to-date information

www.who.int/en/
> The World Health Organization (WHO) is an excellent source of both historical and accurate up-to-date information on many diseases.

Index